EVOLUTION

How a Theory
Made a Monkey
Out of
MAN

SAM McCORMICK

God Bless,
Col. 2:8

CrossHouse

Sam McCormick

Published by
CrossHouse Publishing
PO Box 461592
Garland, TX 75046-1592

Copyright Sam McCormick 2009
All Rights Reserved

Printed in the United States of America
by Lightning Source, LaVergne, TN
Cover design by Dennis Davidson
Cover artwork by Lynn Kaatz

Unless otherwise indicated, all Scripture taken from the Holy Bible,
New International Version, copyright 1973, 1978, 1984
by International Bible Society

ISBN 978-1-934749-69-2
Library of Congress Control Number: 2009942872

TO ORDER ADDITIONAL COPIES FOR $19.95 EACH
(ADD $3.00 SHIPPINGFOR FIRST BOOK,
$1.00 FOR EACH ADDITIONAL BOOK) CONTACT
CROSSHOUSE PUBLISHING
PO BOX 461592
GARLAND, TX 75046-1592
www.crosshousepublishing.com
877-212-0933 (toll free)

Contents

PREFIX 5
CHAPTER 1: EVOLUTIONAL PHILOSOPHY 9
CHAPTER 2: IN THE BEGINNING 15
CHAPTER 3: ELOHIYM: CREATOR OF HEAVEN AND EARTH 27
CHAPTER 4: THE CREATION DEBATE 43
CHAPTER 5: CHRIST AND CREATION..61
CHAPTER 6: CHARLES DARWIN: THE RELATIONSHIP OF UNITARIANISM, NATURAL SCIENCE
AND "ORIGIN OF THE SPECIES"...73
CHAPTER 7: NATURAL LAW 91
CHAPTER 8: IT TAKES A BABY 113
CHAPTER 9: FOSSILS AND THE GEOLOGIC COLUMN 119
CHAPTER 10: THE CAMBRIAN EXPLOSION 145
CHAPTER 11: THE LIVING CELL: ATOMS, MOLECULES AND DNA ...149
CHAPTER 12: CARBON AND RADIOMETRIC DATING 167
CHAPTER 13: THE EVOLUTIONARY POINT OF VIEW 194
CHAPTER 14: BIBLICAL HISTORY: THE AGE OF THE EARTH
AND THE FLOOD 216
CHAPTER 15: ARGUMENTS OF LAST RESORT 242
CHAPTER 16: THE COSMOLOGICAL SEARCH 250
CHAPTER 17: APE TO MAN: THE STILL-MISSING LINK 260

CHAPTER 18: EVOLUTION FAILS TO ANSWER 286
CHAPTER 19: A FINAL THOUGHT 294
HANNAH'S EYES 299
GLOSSARY 303
ACKNOWLEDGMENTS 315
BIBLIOGRAPHY 317

Prefix

As I sat across the desk from Pastor Ron I could hardly contain the excitement I felt over my newfound discovery. It seems a former high-school science teacher, Dr. Kent Hovind, had developed a series of tapes about creation by the hand of God, which devastated the theory of evolution. A friend had given my wife a copy, and I skeptically agreed to watch one of the homemade DVDs.

Coincidentally, Ron shared with me a book he had recently received that outlined how 50 scientists in various fields of study also rejected evolution, based on their own scientific research.

That same day I ordered "In Six Days" and began what has become a three-year search and study for the truth about evolution.

I, like many Christians, had accepted evolution as a proven fact and had found a place for its acceptance in my own belief system. Though an acceptance of evolution had altered much of my childhood understanding of the inerrancy of scripture, I managed to adjust scripture to comply with what was taught as modern science.

What follows, in the pages of this book, is a summary of facts and information gathered from many, many sources. It would have taken me several lifetimes to research everything that is included in this book; therefore, I am indebted to others for their research and willingness to contradict what many people have accepted as sacred evolutional truth. I have also made every effort to simplify some of the more complex information and, as much as possible, explain things in layman's language.

I give credit to the knowledge and research of many in the pages that follow. What I have learned over these past three years would not have been possible without their expertise.

Finally, I am grateful to my wife who not only accepted the task of typing my scribbled notes, but also inspired me each step of the way—from the first DVD she asked me to watch until this project was completed.

What I Discovered

Surprisingly, I discovered a great deal of counter-evolutional writings. Some concentrated on theological and philosophical disagreements with evolution while others dealt with scientific contradictions and illogical conclusions from natural observation. Admittedly, I felt like an eighth-grade student in his first biology class when reading the scientific disagreements, but I forced myself to read and re-read until I felt comfortable enough to share what I thought to be important information.

You can't imagine how many times I asked myself, "How could I have been so wrong for so long?" The truth is that my faith in the awesome power of God was weakened by my own prideful ego. Sadly, my ego provided me a way to justify some of my own selfishness and excuse my lack of faith with my own interpretation of God's Word.

I believe every child of God walks his own path of faith and surrender to God's Son. Your journey, perhaps, is different from mine, and that is how God intends it to be. I also now believe that God's inspired account of creation is the plumb line of biblical inerrancy. If you disagree, please challenge that statement and read on.

Chapter 1

Evolutional Philosophy

Evolution is a logical byproduct of the Socratic philosophy of latent potentials. While claiming the higher ground of empirical science, more often than not, evolutionists are forced to acquiesce to the lower ledge of "yet undiscovered" and/or "not yet realized" explanations. Such a concept should be an anathema (detestable) to a devout evolutional scientist who takes prides on scientific observation and logical conclusions. Many go to great lengths in order to develop what is a purely reasoned and self-contained system of creation by chance. The problem is that many of their theories are launched from a metaphysical foundation; though most devoutly refuse to admit a supernatural component. If evolutionists ever began a sentence with "it is possible that" or "some believe", then their claims would not be so divisive.

Science is validated knowledge, not theory or speculation. Validation comes from repeated measurement and observation. Scientific theory must allow for the possibility of being refuted. Evolutional science would be far better stated in an Aristotelian format: *A will always result in B, may sometimes result in B, will never result in B or may sometimes not result in B*. In other words, theory should always allow for alternatives.

Medical pharmacists know that their miracle drugs may not always result in a miracle; hence, warning labels. If evolutionists would acknowledge that chance generation from non-life may not result in life, at least this theory might come with a warning label. Potential does not reveal truth or cause; it only suggests the possibility of fulfillment (actualization). According to Aristotle (322 BC), matter is only the visible. It is the essence of matter, its purpose and reason for existence, that ultimately reveals its cause—not simply the fact that matter exists.

Socrates believed that eternal substance contained the potential for life, which lay dormant until it was realized. This concept is fairly consistent with "panspermia" (seeds of life scattered by alien beings), which Francis Crick at one time attempted to make the foundation of his belief in the origin of human consciousness. Crick eventually realized he had wandered into the field of metaphysical philosophy and ultimately admitted human consciousness could not be physically determined.

Socrates (399 BC) believed that truth could be obtained by eliminating alternatives resulting in ideas developed but not visible. While certain ideas may never be fully explained, that does not negate their reality, according to Socrates.

Evolution is accepted as reality by many because the one alternative (supernatural creation), in their mind, has been eliminated; hence, evolution becomes reality by default.

My favorite seminary professor by far was Dr. Eric Rust, a professor of philosophy. One of the two most important concepts I learned from Dr. Rust seemed almost incidental at the time he shared them; however, both serve me now to philosophically explain why so many are able to justify their belief in evolution.

Dr. Rust said that every theory, every philosophy and every religion is based on a foundational belief from which a system of tenets and/or consequences will then be built (the details or building blocks).

If one's foundational belief is accepted then, based on sound reasoning and logical assembly of ideas, what will be constructed upon that foundation will also be acceptable; conversely, the opposite is also true. If one's foundational belief is flawed then all that follows will be flawed, even though it is logically consistent with one's foundational belief.

Divine creationists build their foundation on an omnipotent God who is able to create from outside the realm of time, space and substance. The philosophical foundation of Darwin's Theory of Evolution is random chance existing within what he conceives as potential, natural circumstances. As a foundation, only one can be accurate.

Logically, the supernatural foundation is more easily constructed upon. The simple observation of creation provides the first logical building block. I see the sun, moon and stars and, as a natural man, recognize that only a supernatural being could have placed them. Secondly the supernatural exists outside the burden of proof and is not restricted by the natural, which brings me to the second important concept I received from Dr. Rust.

We humans exist in a three-dimensional time, space and substance world—God does not. God cannot be proven by measurements of length, width or depth based on natural time, space and substance. He can be recognized, rationalized and theorized, but He cannot be empirically proven.

Creation by chance evolution is based on a purely physical time, space and substance foundation. Though we may soon discover that the details of evolution often require unnatural phenomenon to provide the blocks for continued construction, any such unnatural occurrences are explained as only brief or sometimes extended periods of time when natural laws are suspended in order for chance, unnatural circumstances, to take place. The point is that evolution is a physical, three-dimensional creation theory requiring natural proof—divine creation is not.

Evolutionists often make the philosophical mistake of trying to disprove the supernatural. Questions like "Why does evil, suffering or free choice exist?" are either philosophical or theological, but they are definitely not physical and do not serve to prove or disprove the supernatural.

The Trap

Divine creationists should not assume that they are required to prove God and thereby disprove evolution. The supernatural cannot be proven by the natural. According to Dr. Rust, God exists in a different dimension, and only He has freedom of passage into the natural.

My Introduction to DNA

If I ever heard Dr. Rust excited over something it was when he lectured about recent DNA discoveries during the time that genetic decoding was still in its infant stages (the late '60s). Dr. Rust saw this discovery as a visual testimony to a Designer's handiwork. He was convinced that man's understanding of

DNA would change the physical and the theological world forever, which it has, physically.

The Bible talks about people with eyes that do not see. No more is that true than in the more recent understanding science has been given through DNA. While man has dramatically advanced physical knowledge, only those with eyes to see have acknowledged the supernatural complexity of DNA and genetic coding.

Those who see only an evolutional similarity built into the DNA of species assume it is one more proof of a common ancestor waiting to evolve into (latent potentials) all the plants and animals of the world.

Those who recognize DNA as evidence of a creator God who has designed creatures to inhabit a common ecosystem recognize, with "eyes to see", that only a power beyond time, space and substance could accomplish such a wonder.

Chapter 2

In the Beginning

In the year of our Lord, 2009, the world observed the 200th birthday of the person who has perhaps created the greatest turmoil in the Christian church since Nero. Coincidentally, the year 2009 is also the 150th anniversary of his book, "Origin of the Species by means of Natural Selection or the Preservation of Favoured Races in the Struggle for Life"—though you will rarely see this politically incorrect full title mentioned in print.

Charles Darwin's theory of accidental creation and evolution of all life forms, including man, created such a boiling pot of emotions within as well as outside of the Christian church that few people today are capable of an unbiased and reasoned evaluation of its claims.

At its core, evolution of the species is hardly the stuff that scientific journals are made of. Nevertheless, because of its direct challenge to the Judeo-Christian faith, many in the scientific community were encouraged to allow its unscientific license of expression. Evolution soon became a household word.

In 1954, President Dwight David Eisenhower instructed the nation's public school system to begin teaching evolution as an empirical explanation for creation. His concern was that

America's school children were falling behind the communist educated children in the fields of science and mathematics.

Since that day in 1954, America's school children have been literally force-fed a theory that rejects the biblical account of God's creation in lieu of a 19th-century theory. On the one hand, children are taught in church that God created the world in six days and that it was good. However, when they arrive at kindergarten they are told that the world was created by accident that through a series of very fortunate events life on earth was also accidentally created millions of years ago and that creation is only as good as man is able to make it.

From that moment on, all that children read in school textbooks, hear on school field trips and are taught in the classroom is that the world happened by chance, that over millions and millions of years life slowly evolved into what one sees today—and look, we have pictures and charts to prove it.

These statements are never challenged because to do so would isolate a child from the intelligent children in the classroom. Evolution, the big bangs, an ancient earth, dinosaurs roaming the earth millions of years ago and cavemen who were once monkeys is all that children are taught. After years and years of indoctrination, what do you think they now believe about God's handiwork?

Coincidentally, President Eisenhower, failing to see the inconsistency, was instrumental in adding the words "under God" to the Pledge of Allegiance and adopting "In God We Trust" as the United States motto, later adding it to the U.S. paper currency. Though history has afforded "Ike" little credit for his contribution to civil rights, President Eisenhower proposed to Congress the Civil Rights Acts of 1957 and 1960.

They served as the most important civil rights legislation since 1870.

One of the more curious aspects of the evolutionary movement is the almost universal acceptance of the theory as fact and worthy of teaching in the public school system. While the early conflict as well as early court cases dealt only with "evolutional science" in conflict with inerrant biblical beliefs, apparently few people questioned whether "evolutional science" was actually validated science at all.

The second curious aspect of the evolutionary movement is how quickly many Christians adjusted their biblical beliefs to accommodate evolution.

Allow me to capsulate the disagreements between man's theory and the biblical account. Later we will discuss each point in detail; however, to begin our study we must first have a general understanding as to what the disagreements are and the significance of the debate.

The Disagreement

1. The theory of creation by evolution is built upon three absolutes that must be true in order for Darwin's theory to be taken seriously.

    A. The earth must be multimillions or billions of years old.
    B. It must be possible for life to begin from non-life by chance, fortuitous circumstances.
    C. Once begun, first-life must include the instructions whereby it is now able to take on nourishment, eliminate waste, survive the elements, replicate

and the thousands of other intricately complex details necessary in order to sustain life. Few of these details Darwin would have been aware of when he presented his theory of slowly evolved life in 1859.

2. The biblical account of creation is built upon one absolute—a powerful God who created the universe, instantaneously created life and continues to sustain it by the power of His Word must exist.

Hence, the true controversy is revealed along with the true starting point for those who propose one view over the other. If there is a Creator God, He is responsible for creation's design. Man's only challenge is to know this god/designer. If there is no god, then it must have happened differently and the theory of evolution is as good as any theory even with all its improvable conclusions.

The Objections

I am well aware that many people will say that God can exist without the Genesis six-day account of creation being literal. That is indeed the position that many Judeo-Christians have adopted.

Without the reliability of God's Word, however, including the biblical account of creation, man is obliged to conjure up his own version of a god. He can mold a god into his own image and create him as powerful or as impotent as he desires. The truth remains, that one's own created god is of no more value than any other god theory or, in fact, no god at all.

In 1964, J.B. Phillips wrote a book entitled "Your God Is Too Small" in which he submits that many people envision a deity no wiser or more powerful than they are. Our challenge should be to understand the broad sweep of the Creator's activity—not to diminish it.

Without a source of divine truth there are no absolute standards and no reason to consider another's good except to receive consideration in return. One may say they wish to retain all the sweet smelling Scripture and discard those Scriptures that fail to emulate their own personal standard of virtue, morality and goodness. One may exclude certain Scripture based on what science or one's own human experience and reasoning can confirm. All of these choices are granted to you by God or by chance. Your decision to accept one over the other is either of no consequence or of major, eternal consequence.

The Significance of This Debate

Ultimately the existence of God as revealed in God's Word is the greater issue—not how God chose to create the world. Is the Bible true? Can I depend on its lessons and promises as truth?

Some will say, "I can believe in God and also believe in evolution." My response to that is simple. You either do not understand evolution or you do not understand God's Word.

How Complicated is This Debate?

I certainly do not want this study to become too complicated for your sake or for mine. Even though the theory of evolution is not able to be tested and demonstrated, some of the evidence that attempts to support or refute evolution may be somewhat involved.

Where Do Scientists Stand?

You may be under the impression that a great many scientists are laboring in the field of evolutional science. That is simply not true. Except for some in the field of natural science, paleontology and archaeology, few scientists bother to investigate the evolutional claims. So naturally, I ask myself the question, why? Why is evolutional science not a priority within the scientific community? Based solely on my own understanding the following are my personal conclusions:

1. For some, evolution, with all its inconsistencies, is the only existing theory of creation that does not include God; therefore, many gladly accept it, without verifiable proof.
2. Some, who recognize the fact that evolution is not a science but rather a theory (sometimes called an assumption) that cannot be substantiated, prefer not to waste their time with investigating something that cannot be scientifically proven.
3. Many who have labored in fields of empirical science (molecular biology, genetics, biochemistry, geophysics, etc.) have discovered that the 19th-century theory of evolution is completely out of touch with the facts of today's science and many of those scientists have chosen not to rock the boat with evidence to the contrary. Thank goodness, many others are willing to speak out and challenge this 150-year old theory based on their own study and faith.
4. Except for a few in the field of paleontological and archeological academia (students and professors),

most natural as well as empirical (that which can be verified and re-verified) scientists have chosen to earn a living outside the realm of college and/or government grants.
5. And finally, evolutionary research does not pay well, especially if you disagree with the concept. In that case, you may lose your job, as many have when their research conflicted with what evolution proposes.

While we are on the subject of scientists, something else exists that must be taken into consideration in this study.

Most scientists, as is true of most professionals, become proficient (even expert) in only one field of endeavor. Because of the amount of time and study necessary to become knowledgeable in a particular field, many professionals concentrate their efforts in only one area of study.

Consequently, as my wife learned from her vocation in stocks, mutual funds and financial investments, lawyers and doctors (as wise as they may seem in their chosen profession) often make terrible investment decisions.

Geologists know rocks; they are not microbiologists. Astronomers observe the vast galaxies of the universe, but you would not invite one of those experts to perform brain surgery.

It is the nature of life and science to specialize, which is a good thing. Therefore, testimony from an expert (a specialist) is very valuable evidence and, as it relates to creation, should be considered along with the testimony of other experts in their specialized field of study. What is interesting about the early proponents of evolution is that few had any training in empirical science and many had a stated bias against Christianity.

Charles Darwin was a divinity student, Charles Lyell was a lawyer, James Hutton was an agriculturalist (farmer) and Thomas Huxley had a moderate education in medicine. None were formerly trained in physics, chemistry, biology, geology or astronomy, though they theorized about each extensively. Of the few early proponents of evolution who actually had a legitimate education in evolutionary studies, a German professor of embryology, Ernst Haeckel, was exposed in 1875 for producing fake drawings of embryos to demonstrate fetal similarities between certain vertebrates (species containing a backbone).

The Average Christian's Silence

A longtime Christian with average biblical understanding is well aware of how those who reject God will use every means possible to slander, confuse, ridicule and embarrass those who trust in creation by God's six-day design. Many are intimidated by those tactics and have chosen to remain silent. It has been made perfectly clear to Christians that if they get too vocal over their beliefs, they will be labeled kooks, malcontents, uneducated, unsophisticated, intolerant, dogmatic, ultra-conservative proselytizers and generally not as wise and intellectually blessed as those who belong to the congregation of evolution.

Even the phrase "designed creation" is a capitulation to the "sophisticated" evolutionists. Christians continue to try in so many ways to persuade non-Christians to come to God by the use of secular reasoning. Rather than say God created the heavens and the earth in six days, many have chosen to say a designer must have created this complex, designed universe.

Sure, there is complex design easily recognized in the universe; however, we as Christians do not worship a designer. We worship an omnipotent creator, loving sustainer and sacrificial redeemer.

When God convicts He never enters the back door of secular science, secular music or secular promises of abundance. When God convicts He pierces one's heart, shatters one's selfish pride and demands one's humble surrender to His perfect will.

The Bible has been and always will be the most read as well as the most controversial book ever written. Written over a period of 1,600 years first in Hebrew and later in Greek, the Bible has undergone hundreds of translations. Based on thousands of original or first-generation manuscripts, it has remained true to its original inscription, as confirmed by many first- and second-century church leaders and manuscripts (copies of the original text).

The Bible was the first book ever printed on the Gutenberg Press (1455 AD, in Latin); and yet, its central character, Jesus Christ, the promised Messiah, according to all we know, only wrote one simple word or words on the ground. As those who sought to carry out the Hebrew law against adultery were about to stone the offender (see John 8:6-8), Jesus began to write in the dirt.

What was it that Jesus wrote on the ground? Was it the seventh commandment, *Thou shall not commit adultery*, or was it a reminder to those in the mob of all the past sins of

Israel or their own personal sins, perhaps, with this same prostitute?

Whatever Jesus wrote in the dirt that day, the angry crowd must have accepted it as divine truth, as, one by one, they dropped their stones, turned and walked away.

Theory Is Not Fact

Finally, many early, even childhood beliefs that sometimes change as we grow older are not always based on new and profound knowledge or discoveries. They rather are the result of small, seemingly innocuous influences.

If someone walked up to you and said, "Do you believe I can turn this rock into a frog?", you would respond "Do I look stupid?" And yet, with enough persuasion, most people have been led to believe that if they wait long enough, and with enough unforeseen circumstances, that is exactly what will happen.

And what if they said that the world as well as all life on earth was accidentally created by violating every natural law in the universe, would you stand in awe of their superior intellect?

Evolution is a theory that says the impossible is possible if given enough time to occur. Don't get caught up in the "what ifs" and "it could be's". Imagination is not science, theory is not fact and speculation often creates bankruptcy.

Early in life we all learned the axiom that two sides exist to every story. Intellectually we understand that truth; in practice, however, we quite often overlook the other side. Our side seems so compelling that to consider another point of view seems a waste of time. But, there is always another side to every story and only one side may be true.

Proof

Neither the Genesis six-day account of creation or Darwin's theory of evolution can be scientifically (empirically) proven—it cannot be repeated in a test tube or a lab.

Evidence

One's belief in God cannot be sustained based only on an intellectual understanding of evidence. Evidence can be manipulated and skewed. Emotion, at times, overrides evidence and creates a framework for faulty decision making. Evidence may reinforce a belief, but evidence alone is not empirical proof.

When an airplane becomes disabled and begins to fall towards the earth, who or what is to blame for the crash? The evidence may indicate that the airplane or the pilot is at fault; however, based on the natural law of gravity, the earth is at fault for pulling the airplane down. Were it not for gravity, the airplane would remain floating above the earth.

Faith and Intellect

Faith in God is a love relationship, not an acknowledgment of evidence that tends to prove one's belief in God. That is not to say that intellect is not involved; nevertheless, intellect and evidence do not create a relationship to God. Only surrender to (trust in) God creates a love relationship to God.

While God desires a relationship of love and trust with man, He also *would not have us ignorant* (1 Thess. 4:13). God has provided all the signs necessary to grant man intellectual awareness. He has not, however, removed man's choice to accept or deny those signs.

With the coming of Christ, God provided His final and ultimate sign (evidence, if you will) of not only His existence, but also His plan and purpose. God's purpose of restoration—restoring a broken love relationship—is clearly seen and fulfilled in Christ.

Evolution is a suspected theory with suspicious evidence. It is the other side of the story. It begins, for many people, with a rejection of God and/or God's inspired Word. Once that foundational disbelief is established, anything that follows as suspected evidence becomes proof, because it must be true if there is no supernatural power to say otherwise.

Intellectually speaking, if God did not create the universe and life on earth, then it must have happened by chance. If man was not fully formed in the beginning, then he must have evolved over long periods of time by the most fortuitous circumstances imaginable. The fact that most, if not all, of those circumstances are contrary to the known laws of nature does not matter—it must be so, which makes it proof for those who reject Elohiym, the Creator of heaven and earth.

Chapter 3

Elohiym: Creator of Heaven and Earth

Christian antagonist Frederick Nietzsche, a late 19th-century German existentialist, said that it is our preference that decides against Christianity—not argument. By this he divided people into two categories—those who choose to accept God in Christ and those who choose not to accept. Nietzsche contends that neither physical evidence nor scientifically verifiable proof will cause a man to accept that which he chooses to reject.

The creation debate is an extension of the existence of God debate. Who would be so naïve as to not recognize complex design in creation? Except for the most hardened atheist, nature's complex design is strong evidence for the existence of a designer.

It was Nietzsche himself, with his lofty self-awareness, who failed to realize that his humanistic philosophy and rejection of Christianity and divine moral truth was itself evidence of moral awareness and, thus, creation not only by design in substance but also design in moral and conscious awareness.

From the beginning of creation—and most scientists now believe creation had a beginning—man has sought to discover his origin (where did I come from?), his identity (what is my purpose?) and his destiny (where am I going?). Invariably each debate evolved into two diametrically opposed preferences—the existence or nonexistence of God.

C.S. Lewis said that man's attempt to seek some type of understanding or happiness apart from God has resulted in "all that we call human history—money, poverty, ambition, war, prostitution, classes, empires, slavery—the long terrible story of man trying to find something other than God which will make him happy."

Human history is inexorably linked to faith and religion. Nations have formed as a result of religious beliefs or the rejection of the beliefs of others. Wars have been waged in the name of religion and the promise of divine justification and protection. Codes of conduct, laws of state and social norms have all emerged from man's moral awareness and subsequent religious convictions.

To deny that moral awareness has not played a primary role in man's human existence is to deny the past 5,000 years of recorded history and 3,600 years of recorded biblical and church history.

Early Christian philosophers proposed that moral awareness, along with conscious awareness and the "irreducibility of complexity" (all components of a system, be it cellular or celestial, are necessary for that system to function), are the three most philosophically logical arguments for the existence of God. It was Darwin himself who questioned the reliability of any moral awareness that evolved from a monkey. Darwin readily admitted that there are questions, including those of moral and conscious awareness, that evolution could not answer.

David Hume, an 18th-century Scottish, naturalistic philosopher, proposed that only those things that can be empirically observed and/or experienced, based on one's five senses, can be considered meaningful. The supernatural, therefore, falls into the realm of meaninglessness.

Later, Immanuel Kant, an 18th-century German philosopher, refuted Hume's early attempt at empirical logic when he suggested that one's five senses only observe and/or experience that which is individually observed by one; and thus, not empirically meaningful for all. Kant did, however, allow for the existence of a god, though he determined that no way existed for man to know his nature.

Aside from all the rhetoric from natural science, philosophy and also religions, the question remains—did God create man or did man create God?

Could Human Consciousness and Moral Awareness Evolve?

Evolution teaches that life was created out of non-life. Non-living atoms and molecules were given life by the introduction of a fortuitous and yet unknown energy source. Thus, order arose out of chaos, intelligence eventually evolved from nonintelligence and morality out of amorality.

The fact that all men do not follow the same moral code is not evidence against innate moral awareness. If moral awareness did not exist, evil men would not seek to excuse its violation. Justice would not be sought or expected. There would be no belief in human rights. Even the willingness to tolerate evil is recognition of the good.

If morality was genetically acquired—which is the only evolutionary way to explain its existence, there would be no absolute moral standards incumbent upon man to observe. While secular as well as religious communities are most often formed based on what is morally acceptable to the group, such communities could never form as each man fought for dominance and self-satisfaction in an evolutionary "survival of the fittest" world.

Some evolutionists suggest that a moral gene eventually mutated into existence out of the need to survive. DNA strands are matter, not metaphysical. They have no consciousness as Francis Crick, the father of DNA coding, ultimately determined. What have evolved are the concepts of freedom, justice, charity and equality, all a result of man's innate moral awareness.

Is the Christian Church Partially Responsible for the Acceptance of Evolution?

You may find it difficult to believe that the Christian church has played a significant role in the general acceptance of the theory of evolution. If so, consider the following observations.

The Greek New Testament refers to the "ecclesia" as the church, the body of Christ, and it must be understood or defined as separate from what one refers to as the institutional church or assembly house. Citizenship in the body of Christ (the church) is by invitation only. It is God who calls and the Holy Spirit who confirms a believer's citizenship in the kingdom of God, the earthly "ecclesia."

That being said let me also say that well-meaning believers, not just those who live on the fringes of the institutional church, have perhaps, unwittingly, advanced the cause of evolution. Allow me to mention four ways in which I believe that to be true.

1. **The "Abandon Ship" Mentality:**
   With the onset of what has been called "biblical criticism", what theologians like Albert Schweitzer, "Quest for the Historical Jesus", and even Karl Barth, who rejected much of the liberal views of Schweitzer, did was to challenge the inerrancy of Scripture. It was not so much the contextual reinterpretation of some Scripture passages that caused some to abandon the conservative inerrancy "ship", but rather the challenge to God's supernatural intervention into creation and human history. Once less supernatural interpretations of Scripture were made acceptable, then all Scripture became suspect—especially creation Scripture. This new view of Scripture was occurring during the same time that evolution science was being introduced as an alternate view of creation.

   What developed over the past 150 years in the church is best described as a reinterpretation of the Genesis six-day creation account into a theory more compatible with Darwin's theory of evolution and an ancient earth. Unfortunately, this reinterpretation has caused many Genesis believers to "abandon ship" and reinterpret their creation belief into what is most often referred to as "evolutional theism" (the belief that God allowed millions of years to pass during each act or day of creation).

2. **The "Please Love Me" Syndrome:**
Modern Christians have so yielded to the world's criticism of intolerance and dogmatism that they have abdicated those Scriptures that seem "unloving" or irrational to the world and chosen to only teach those beliefs that are consistent with the world's secular opinions. What many fail to recognize is that with each step the Christian church retreats into charming the world, the world erases more and more of those foundational beliefs crucial to the Christian faith and the Christian walk.

What Scripture teaches concerning the world's acceptance is exactly opposite to what many churches now strive to achieve. While "love your neighbor" is both an Old and New Testament command, some now operate under the delusion that they can "entice" as a way of loving people into the kingdom of God by removing all the disagreeable passages of Scripture. They have sought to persuade the world by conforming to the world. They simply do not believe it must be as Christ said, *"all men will hate you because of me"* (Luke 21:17) and, *"if the world hates you, keep in mind that it hated me first. If you belonged to the world it would love you as its own . . . If they persecuted me, they will persecute you also"* (portions of John 15:18-20).

As a follower of Christ, one does not seek the love of the world; he or she seeks the will of the world's Creator. It is the willingness of Christians to compromise belief for the sake of worldly acceptance that enabled the theory of evolution as well as other secular persuasions to gain acceptance in the church.

3. **The "I Am Intelligent, Too" Disclaimer:**
By allowing science to determine one's view of Scripture, the church has apologetically said to the world, "Don't make fun of me, even intelligent people can still believe in the God of creation." As hard as one might try, the spiritual cannot be adjusted to the material in order to satisfy those who choose to reject the spiritual. Nietzsche was correct—it is not argument that makes a person decide for or against Christianity, it is preference (the will).

A Christian belief system adjusted to man's human wisdom does not encourage acceptance, it reinforces rejection. The Gospel will always appear as *foolishness to them who are perishing* (1 Cor. 1:18) just as in the days of Noah; and yet, many who represent the Christian church continue to water down gospel truth in order to achieve intellectual acceptance from the perishing.

A little chorus I now recall from my childhood says it all. "I stand alone on the Word of God, The B-I-B-L-E." Christian believers may indeed appear foolish to the world; but only until that day when Christ returns to redeem His own.

4. **The "Why Can't We All Just Get Along" Philosophy:**
Finally, and perhaps the most damaging cause for the advancement of the theory of evolution in the Christian church is the humanistic, New Age belief that many paths to God exist. Such a belief undermines the

entire Old and New Testament promise and fulfillment of redemption in Christ.

Once Christ, "the Word become flesh", is removed as the cornerstone of all creation, the beginning and the end, then biblical truth is no longer the source of light to the universe. Any belief, any path or any theory becomes acceptable to the world. Knowingly or unknowingly some in the church have allowed human understanding to replace biblical truth. The theory of evolution may never have gained traction had not Christians been so anxious to accommodate what was acclaimed as creation science in order to get along with the world.

Some men will choose to reject God and follow a false belief system. As difficult as that is for a believer to accept, it is, nevertheless, what Christ taught. In spite of that, the church cannot and must not allow itself to accommodate a weak, allegedly scientific yet logically questionable theory of creation instead of "standing alone on the Word of God."

Sadly, I must confess that during my own spiritual pilgrimage I also fell into each one of the four church failures. Perhaps that is all the proof I need to acknowledge the existence of Satan. Doubt, confusion, half-truths and pride are his greatest tools. Even a strict Bible conservative, committed to the pastoral ministry, can be led to compromise his faith in biblical truth.

It is not that we as Christians do not and should not continue to grow in knowledge and understanding of biblical truth through greater linguistic and contextual interpretation of scripture; it is rather that times exist when we move too quickly to abandon the clear message of the Bible when it is challenged by the world.

As Christ is the cornerstone of the church, it is now my firm belief that the Genesis biblical account of creation has become the cornerstone of man's understanding of all biblical truth. Many will reject what I have just said as they are so deeply indoctrinated in the evolutionary theory. They will continue to trust in a theory of their own preference; however, once one begins putting scissors to the Scriptures, I ask you, how reliable is what's left going to be?

Clarifications . . .

1. Moral awareness is not the same as biblical truth. While the teachings of the Bible are considered by Christian believers to be ultimate moral truth (what is right before God), moral awareness is simply the conscious awareness of a right and wrong. Nietzsche did not reject moral awareness; he rather taught that morality was relative to the situation—that there was no such thing as ultimate or absolute truth or an absolute truth-giver.

2. Conscious awareness is man's ability to reason, not simply to react from instinct. It is not that man never reacts from instinct, but that man is able to employ conscious reasoning to control or direct instinct. Man's responses can be conditioned with the proper stimuli; however, man has the ability to change his response by reasoning a more appropriate or less costly reaction.

Anyone who has ever watched a dog in a fenced yard try to escape is aware that with the proper incentive a dog will go through what is called "trial and error" in order to achieve its goal of escape.

"Trial and error" is not conscious awareness; not unless your dog is first able to stand back and reason whether climbing over, digging under or finding a hole in the fence is the most expedient method of escape.

Evolutionists would suggest that instinct, conditioned responses or "trial and error" are only one gene away from conscious awareness. The question we will soon address is whether one gene or Darwin's "survival of the fittest" is scientifically sufficient to explain man's conscious awareness.

3. The "irreducibility of complexity" means that to alter one amino acid from its present position in a chromosome strand or to remove one molecule from a cellular DNA ladder or to slightly alter the earth's gravity or rotation would destroy the entire physical, cellular or celestial system. Our bodies, our ecology and our universe are so finely tuned that all parts must work and continue working in proper balance and sequence in order for any of the parts to function properly and survive.

Plants cannot live without the carbon dioxide provided by animals, and animals cannot survive apart from the oxygen provided by plants. The earth's oxygen level is 21 percent of our atmosphere. If it were 25 percent, fire would destroy all life. If it were 15 percent, human beings would suffocate. The relationship of the earth's weight and thickness maintains the proper pressure necessary to hold back earthquakes and allows access to, as well as control over, the release of oil and gas. All solar systems operate under the proper degree of gravity and centrifugal force in order to maintain their orbit and not collapse in on themselves.

In each case, critical complexity is required in order for any and all to function—complexity that cannot be reduced.

4. Observation does not prove origin, though it can demonstrate complexity, codependency, function and even mathematical probability. Natural law does not produce complexity; it rather sets the boundaries in which the origin of life and life complexities occur. Life is more than the existence of chemicals. It is the complex codependence of chemicals and natural law that allow for life and the survival of life.

The foundation of this study is logic, evidence and faith. The evidence is abundant and, perhaps, at times tedious. At least four absolute distinctions exist between evolution and the Genesis creation account to which both sides agree:

1. One claims to be scientific and the other does not.
2. One claims to be proven and the other does not.

3. One claims to be faith-based and the other does not.
4. And, finally, one relies on the biblical testimony and the other does not.

The burden of proof lies solely on the shoulders of evolutionists. Creationists claim no proof. God and the supernatural cannot be scientifically proven; consequently, it is up to evolutionists to disprove Genesis and, thereby, cast doubt on the God of Genesis.

Faith

Faith is not substance. It cannot be put in a test tube or under a microscope. Faith is, according to the Bible, trust and confidence in that which cannot be seen. However, and that is a huge however, faith is not devoid of logic and evidence. There is no such thing as biblical, blind faith. The Bible says that the heavens declare the glory of God; man must simply open his eyes. Behind any theory (faith in a hypothesis) there is some degree of faith based on logic, evidence and or observation.

When Jesus spoke concerning the faith of a child, He elevated the nature of believer's faith. While one's "adult" intellect is an intrinsic part of believer's faith, Christ desires a faith that compels "child-like" surrender to God, not simply intellectual, adult acknowledgment of God.

Soon we will examine the evidence from the theory of evolution. Some believe that the evidence is strong enough for them to have faith in evolution. Some will say that the evidence is so overwhelming that to them evolution has been proven. That, of course, is impossible. Neither a six-day instantaneous Genesis creation, nor a four-and-a-half billion-year evolutionary creation can be scientifically proven.

Evidence and logic are available to both sides to support their confidence; however, faith will ultimately determine one's belief (when one surrenders his trust) no matter on which side of the debate one falls.

There is an additional ingredient in one's spiritual faith beyond logic and evidence, and that is the presence of God's Spirit. You will feel His presence even if you reject His existence. You cannot escape being a spiritual creation. For me to tell you that God made you that way may mean little to you. But in those moments when you think you are alone in your thoughts, and you become aware of a presence beyond yourself, you question whether God is real and it's not so easy being an atheist.

So Much Evidence – So Little Time

I truly wish that the theory of evolution could be dismissed with one or two obvious contradictions, and if Darwin's theory was written last year, that might be possible. But because we are dealing with 150 years of acceptance by the vast majority of people, it will take more than a couple contradictions to alter the existing belief in evolution; even though all it should take is one.

Because creation cannot be repeated, tested and retested, it cannot be proven; therefore, like a criminal prosecutor without an eye witness, our case "beyond reasonable doubt" for instantaneous, designed creation must be built on an abundance of circumstantial evidence.

Not counting falling asleep, I ask only two things of the Christian reader:

1. Don't say "I just accept it by faith." Spend time learning some facts in order to confirm your faith. Study to show yourself approved by God.
2. Learn why you believe what you believe. Use whatever resources you have, including this study, to build a firm foundation to your belief. Read, reread, examine and question those things the world as well as those the Bible are saying. It's not a sin to ask questions about God's Word. It's a sin to reject God's Word.

Choice

Every writer who has ever attempted a paper on creation or a related article on creation begins with a foundational belief. With creation of the universe (substance) only three possible foundational beliefs exist, though the details may differ:

1. The world was created by chance circumstances.
2. The world was supernaturally created and designed.
3. The world has always existed.

Even if life was later seeded by aliens, that does not address the issue of creation of the universe (substance and matter) in which to plant those seeds, and that does not change those three foundational options.

I have never been more convinced, having personally studied creation by design or by chance over the past three years, that one's foundational creation belief also affects one's moral and spiritual beliefs. Like the cars on a train, the creation engine, with all its baggage and passenger cars in the middle, determines one's judgments on all of life until life is terminated (the caboose)—the end of the train.

Some will say, as many have, that the theory of evolution has destroyed the faith of countless young people once they encountered the alternative to God's divine creation formula. I respectfully disagree.

It is not the theory of evolution that destroys spiritual faith; it is rather man's human pride, which demands that all divine sovereignty bend in favor of one's own human understanding. When truth is the creation of man, then virtue, morals and even God are relative to one's human truth.

Evolution is an excuse not a cause. George Lyell, the creator of the textbook "Geologic Column", was delirious with joy when he found an excuse to reject the God of Genesis. Countless others use evolution to free themselves from divine commands and righteous living. Others use evolution to reject those biblical texts they find objectionable. The Mormon Church used evolution to prove that the black race was inferior to the white man, and many Christians used evolution to promote racist attitudes long held in the white community.

And it is as God planned it. He created mankind with free choice, and if a man chooses to determine his own beliefs, reject God's Word and God's Sovereignty, any excuse will do.

Chapter 4

The Creation Debate

The theory of evolution, as proposed by Charles Darwin in 1859, sets about to redefine the origin of life and its theoretical evolvement towards life as it is observed today.

Not all those who initially gravitated to a theory that suggested life evolved from non-life and continued through natural selection, mutations and extraordinary environmental conditions did so with the intention of denying the existence of God. Though Charles Darwin himself expressed grave reservations on the subject of God and religion in general, his observations and subsequent cataloging of various plants and animals led him to conclude that life evolved from a common ancestor and not as the biblical record portrayed—kinds of plants and animals created fully developed and independent of one another. Since Darwin, however, a great debate has arisen as Bible believers and disbelievers challenge one another primarily over the reliability of faith verses the reliability of science.

But is that truly the issue? As you enter into this study, hopefully, you will begin to understand that the debate has been wrongly labeled. True science and the biblical testimony need not and should not contradict one another.

Neither science nor biblical faith should be threatened by the claims of a theory, any theory. As the Bible says, *Faith is being sure of what we hope for and certain of what we do not see* (Heb. 11:1). Paul, in his second letter to the church at Corinth, said: *We fix our eyes not on what is seen, but on what is unseen. For what is seen is temporary but what is unseen is eternal (2 Cor. 4:18).* The writer of Hebrews also said, *By faith we understand that the universe was formed at God's command, so that what is seen was not created out of what was visible* (Heb. 11:3).

Biblical faith is based on an eternal, unseen God who created a temporary, visual universe out of that which is unseen. To the Christian, that unseen power of creation is understood to be God's Word.

Science, on the other hand, is bound by temporal (time, space and substance) constraints. Science must function in the realm of observable substance, experimentation, measurements as well as questions from antagonists. And, thank goodness it must. The next time I board an airplane I want to feel confident that the science of aeronautics has been properly adhered to in the construction of that flying machine. I want its design as well as its mechanical components to have been tested and retested, measured and measured again. It is the scientific process and the seriousness in which scientists approach their individual field of scientific mythology that has enabled mankind to achieve remarkable accomplishments. If scientists remain true to the scientific approach, we can all have reasoned faith in its assertions.

True Science Should Encourage Faith . . .
Faith Always Encourages Science

Where would science be without faith? What is a scientific theory, if not conceptual faith in possible scientific fact? Faith, however, does not create scientific fact.

In the early years of aeronautical engineering certain theorists conceived of flying designs that only they had sufficient faith to test. In fact, many of those designs were tested only once. Most mentally competent people would not have enough faith to jump off a cliff wearing a pair of feathered wings. The hope of flying may have sounded plausible in theory, however, the testing quickly proved otherwise. How were those winged humans to know that two of the reason birds can fly are that their bone mass is full of air pockets making them basically hollow and that it is not just the feathers that enable flight but rather the construction and aeronautical design of a feather.

On every feather a system of barbs and tiny barbules exists that requires a microscope to be seen. The barbules on one side of the barbs are ridged and on the other side are hooked. The barbs on each feather become connected as the hooked barbules from one barb connect with ridged barbules from another adjoining barb thus allowing for flexibility and flight.

The Value of Doubt

Doubt is often that which gives wings to faith and enables a believer to overcome doubt. Indeed, doubt may ultimately confirm faith. Science is also driven by doubt. Who knows how many injuries were averted when doubt caused the Wright brothers to be more cautious while testing early aeronautic designs. It is doubt that encourages a research pharmacist to first test their drugs on rats rather than humans.

While correctly applied scientific experimentation will certainly relieve doubt, even so, a wise person will take precautions when testing a theory.

Is Evolution Proven?

It took perhaps 100 of the 150 years since the theory of evolution was presented by Darwin in "Origin of the Species" for society as a whole to grant this theory "proven" status. No test was taken and no experiment took place. Findings were not measured—not even once. Oh, there was evidence, some of it quite compelling.

For many, the evidence was sufficient to convince them that the theory was fact, and so it became fact. Newspapers acknowledged it as fact. Magazines and school textbooks said it was a proven fact. Some school administrators were so sure it was fact that they forbid teachers from mentioning the alternative theory. Museums and other publically funded institutions said it was proven, and that was that. We were all reluctant to question science. We had faith in science.

The problem, as we are now discovering, was that few people bothered to separate scientific fact from an assumed theory, even a compelling theory. Over the years that followed this theory gained traction; until today, it has a life of its own.

Initially most people who accepted this theory as fact did not reject God in the process; they rather altered their view of creation as it was presented in the Bible. Evolutional theists accepted the theory and rationalized their rejection of the biblical record as God choosing to use Darwin's evolutional theory and the passage of time to accomplish His creation. Who would have imagined that God would be so impressed with Charles Darwin or that it took Darwin to finally reveal God's creation formula?

Choices and Conclusions

As I approached this study I was made aware that many of my Christian friends believe in creation by evolution. Many pastors and Bible professors believe in creation by evolution in one form or another. I am told that the Pope believes in creation by evolution. The majority of people in the world believe in creation by evolution. How could they not? It has been proven.

> Following his retirement and wishing to find something to fill his days, an elderly gentleman took a job at the city museum and was assigned a security position in the wing that held various fossils as well as a partially erected skeletal display of a dinosaur—a tyrannosaurus Rex to be exact. He enjoyed talking to the many school children who visited the museum.
>
> One day a teacher leading one of the school groups sought to engage the elderly gentleman into her class lesson, so she asked him, "The children want to know how old are these skeletal remains"? "Madam", he said, "Those skeletal remains are 60 million, three years and two months old." To which she replied, "Well, that's very exact. How can you be so precise"? He replied, "When I came to work at this museum, I was told that these remains were 60 million years old. Since that day I have worked here three years and two months, not counting this week." (Copied)

The point is that many of us accept that which we are told is scientific fact and pass it on as truth. Few of us have the scientific background or the time to investigate those assertions and, with an open mind, draw our own conclusions.

Much of the control that science and government maintain over people comes from one's failure to investigate the alternatives and consider the evidence. We take far too much for granted.

If the pill bottle says, "take two every six hours for relief", we take two every six hours. If the package says, "FDA inspected" we buy it, cook it and eat it. What choice do we have? We are not scientists.

Actually, we have many choices, but they require an open mind and the willingness to spend time in understanding the explanations, choices and alternative conclusions. Our school system has denied school children the opportunity to even hear the alternative to evolution. They reason that the alternative promotes religion, God forbid, and that according to the ACLU and those who prefer a godless society, it is contrary to the U.S. Constitution. That of course is hogwash.

Rather than argue the constitutional issue, instead, consider more closely what is being said.

1. The only alternative to evolution is creation by design, which implies a designer. If there is a designer, he must be a god. Only a god could construct such an amazing and complex universe.
2. To deny an alternative to chance creation eliminates at the same time one's consideration of and response to God, the creator/designer.

Even the ACLU recognizes that one's decision concerning who or what created the universe is the battleground for the hearts of men. Sure, you can continue to believe in God while trusting in evolution, but why? Everything we know about who God is and what He desires comes from the Judeo-Christian biblical record. If the biblical record is untrustworthy, if science can disprove it, why would a person have faith in its promises?

You may say that except for the formula for creation, I will trust the Bible, and that is exactly the road that many Christians have traveled. Unfortunately, once the biblical record of creation is called into question, so too must the miracles, the teachings and anything else that do not fit one's human understanding. And man, therefore, sets about to rewrite Scripture in man's own image.

C.S. Lewis remarked that most of those who claim to be able to read between the lines of Scripture have lost their ability to read the lines themselves.

On the subject of evolution, Lewis wrote, "This myth is the imaginative and not the logical result of what is vaguely called modern science. It is based on emotional needs. It treats data based on selection of scientific theories and it uses great freedom in selecting, slurring, expurgating (to amend by removing offensive or objectionable matter) and adding at will."

What's the Damage?

While Darwin's evolutionary theory has provided a possible creation scenario for those who honestly as well as logically seek a scientific understanding of creation, it has also given aid and comfort to hate-filled proponents. Adolf Hitler took advantage of the ape-to-man theory by exterminating millions of Jews claiming them to be more ape than man.

His excuse when Jesse Owens defeated his highly acclaimed Aryan track stars was that blacks were near apes, which gave them a physical advantage over men more human. Many including George Lyell, one of several who inspired Darwin, have heralded the evolutionary theory as proof that there is no god. Their persuasive arguments have convinced and continue to convince many. When one determines there is no God, God loses a child and the child loses his eternal Father.

Without God's authority to instruct the ways of man, humanity is left to determine its own moral and ethical authority, and that has always been a formula for disaster.

Karl Marx was raised as a Christian but after studying evolution he sought to "dethrone God" and destroy capitalism. When he died, six people attended his funeral. For those who believe that a New Age philosophy or progressive theology will bridge the gap between theism and Gnosticism (trust in the knowledge of man to provide goodness and humanity's upward moral progression), I have sad news. It does not work, it has never worked and it will never work. God is the source of goodness and knowledge, not man.

Just as evolution is a theory of death, where one life must die allowing for the survival of a mutation, so too, with the death of God in the hearts of His creation, will come the death of God's authority and the birth of a godless mutation. We are only now beginning to see the face of that mutation. If you believe the world is getting kinder and gentler without the authority of God, you are living in a dream world.

Our school system is perhaps one or two generations away from complete chaos and our prison system is at maximum capacity. Hate and greed is the order of the day, not compassion and selflessness.

Since the introduction of evolution in America's public school system and the expulsion of prayer and Christian-oriented activities, violent crime is up 995 percent, the unwed birthrate is up 325 percent and unwed pregnancies among girls of ages 10-14 are up 553 percent. The divorce rate is up, child abuse is up 2,300 percent and the use of illegal drugs among youths is up 6,000 percent.*

*Today's figures would be higher as these percentages do not represent the past 10 years between 2000 and 2009.

Those of us who surround ourselves with nice people in nice church buildings and nice communities are perhaps blinded to the downward spiral taking place in our country and our world. As the world's population increases and the Judeo-Christian believers decrease, the inherent selfishness of man will inevitably consume whatever tendency man may now possess towards altruism and "love thy neighbor." Remember the book and later movie, "Lord of the Flies" by William Golding? Anarchy does not inspire benevolence.

Honest politicians are rarely elected. Self-interest groups determine our laws, our energy sources and soon, the food we will be allowed to grow and eat. Broken families, broken lives and broken children reflect what's happening in the world today. This is what happens when godly "mutates" into godless.

Though the secular effects of a belief in creation by evolution are beginning to be witnessed today, for those who have eyes to see, the spiritual damage began decades ago.

Ann Coulter, in her book "Godless" wrote concerning evolutionists, "These aren't chalk-covered scientists toiling away with their test tubes and Bunsen burners. They are religious fanatics for whom evolution must be true, and any evidence to the contrary—including, for example, the entire fossil record—is something that must be explained away with a fanciful excuse, like 'our evidence didn't fossilize'."

Remember, faith is not in conflict with logic or valid scientific study. Faith in God's Word encourages scientific study and greater understanding of our world.

The Old Testament writer of Proverbs wrote,

> *Get wisdom, get understanding, do not forget my words or swerve from them. Do not forsake wisdom and she will protect you, love her and she will watch over you. Wisdom is supreme; therefore, get wisdom. Though it cost all you have, get understanding* (Prov. 4:5-7).

The writer of Proverbs also wrote, *Pride only breeds quarrels, but wisdom is found in those who take advice* (Prov. 13:10). In other words, it is pride that keeps a man from considering evidence and mature advice. True wisdom is granted to those who seek full knowledge, thereby allowing them the understanding with which to judge evidence.

Ultimately, however, wisdom reaches a point where only faith may proceed. Both science and religion will eventually reach that point.

The Incredible Alternative

Evolutionist professor of zoology, D.M.S. Watson said that evolution is accepted because of "the collapse of alternative explanations." In 1927 he stated, "evolution is accepted not because it has been observed to occur or can be proven by logical coherent evidence to be true, but because the only alternative (creation by design) is clearly incredible." He admits that "while the fact of evolution is accepted by every biologist, the mode in which it occurred and the mechanism by which it has been brought about are still disputable."

Four Quick Observations:

1. Every biologist does not accept evolution in fact or theory.
2. If the criterion for dismissing an alternative explanation for a scientific solution is incredible, then the Salk vaccine, formulated out of weakened strands of the polio virus itself, would have never been discovered. At the time, how incredible was that?
3. A defense lawyer would be thrilled to try a case when the accused was being prosecuted on the basis of "lack of evidence."
4. Finally, to suggest that there are many possible alternatives other than "chance generation" and subsequent evolution by chance upward mutations; and, on the other hand, creation by design (God) is blatantly dishonest. Once you exclude alien migration, only those two alternatives remain.

The debate is not between many alternatives. It is between the only two life creation alternatives.

Once again, C.S. Lewis said, "Logic cannot be alien to the nature of the universe—nor can the universe be alien to logic."

What you are being asked to do in this study is to logically consider the evidence. While many Judeo-Christians are so grounded in their faith that no logic or evidence could possibly persuade them to believe anything apart from what their faith has determined to be true, others are not so grounded. Some, but not all, of the evidence is circumstantial. Based on one's bias, a person may easily be persuaded into one opinion or the other. Nevertheless, there is a great deal of natural, celestial as well as microscopic evidence, all of which should be taken into consideration in order to make a logical and, not to be excluded, God-directed decision.

Agendas

Strictly speaking an "agenda" is a plan or outline of things to be done—a format. As an agenda relates to creation, it should be defined as a predetermined mental or emotional plan whereby all related evidence and/or information is selected, considered, advanced or discarded based on the boundaries of that agenda. In simpler words, if the evidence conflicts with one's desired agenda it will generally be denied, discarded or marginalized.

There are many subjects in which one has no agenda. They have no predetermined understanding or emotional involvement—"they have no dog in that fight." On the subject of creation, however, most people do have either mental or emotional agendas from which it is difficult to separate themselves.

Mental agendas based on the body of knowledge one presently possesses are the easiest to alter as new evidence or knowledge is acquired. Emotional agendas are more difficult to alter as they are based on desire and hope rather than evidence and logic.

Emotional agendas are not intrinsically bad. They do sometimes blur the picture. They limit the proper application of knowledge and logic. Creationists, those who insist on a divine design are not free of an emotional agenda. However, that agenda, hopefully, is determined by biblical faith that filters knowledge.

Evolutionists have agendas as well. Some of those agendas are honest and based on a logical pursuit of knowledge. Those evolutionists can be persuaded to alter their beliefs, as many have, based on additional information and/or logic. Those agendas are more mental than emotional.

It is the emotional evolutionists who will have the more difficult time being persuaded by the evidence to change their belief. They are committed to a desired agenda. One writer suggests that some of those who gravitate to Darwin do so in order to make atheism respectable and thereby, reject biblical accountability in order to justify a chosen lifestyle. Perhaps that is true. However, eternal or ultimate truth is not relative. To deny truth in the hope of gaining acceptability is ultimately a dead-end street, and it does not change truth.

As a Christian, sometimes my agenda is to persuade, but not by emotions, which are often transitory and erratic. I leave it up to God to convict; I have no reservation against seeking to persuade someone to understand and accept the biblical account of creation and redemption.

In my opinion, neither the honest evolutionist nor the honest creationist are eggheads (the closest definition of "egghead" I found in the dictionary was "eggplant" which is defined as a table vegetable).

Because all agree that creation cannot be replicated in a test tube, aside from the biblical record, both alternatives rest upon (1) present physical observations and the evidence they provide, (2) inferences from observations, and (3) logic to decipher evidence even though it may conflict with either one's agenda or initial belief. Neither evolution nor design creation should be threatened by the evidence. Honest men who desire to know the truth will do so with an open mind and an open heart.

The Debate Redefined

While some Christians believe they can alter the debate by assuming positions that acknowledge an evolutionary theory as well as a creator god, in my opinion—and I believe logically speaking—that is impossible. Not only does the evidence we will soon discuss prohibit such a position, it is also unimaginable that with creation of the universe in Christ and ultimately creation of man, that God would leave anything to random chance over long periods of time.

Correctly stated, the creation of life debate is between the only two logically feasible, physically tenable and/or spiritually acceptable alternatives:

1. Creation by the Word of God . . .
   or
2. Creation by chance generation and natural selection over long periods of time.

Finally, allow me to humbly address two questions to those of you who consider yourselves people of faith.

1. Is the weight of your faith based more on hopeful feelings rather than unwavering conviction?

   There is a difference you know. Paul said, "For I am convinced that neither death nor life, neither angels nor demons, neither the present nor the future, nor any powers, neither height nor depth, nor anything else in all creation will be able to separate us from the love of God that is in Christ Jesus our Lord" (Rom. 8:38-39). Paul was later beheaded in Rome.

   The early Christian overseer, Ignatius of Antioch (ca. A.D. 35-110), believed to be a student of the disciple John, was arrested in Syria and transported to Rome, later to be martyred in the Flavian amphitheater. Ignatius wrote to the Christian church in Rome these words: "I am dying willingly for God's sake, if only you do not prevent it. I beg you; do not do me an untimely kindness. Allow me to be eaten by a beast, which is my way of reaching to God . . ." and it was so.

   Polycarp (ca A.D. 69-155), overseer of the church in Smyrna, was arrested for being a Christian and told to proclaim "Caesar as Lord." In the midst of an angry mob Polycarp responded, "Eighty-six years I have served Christ, and He never did me any wrong. How can I blaspheme my King who saved me?" He was taken to a stake and burned to death.

2. Would your faith in Christ be stronger if you trusted the literal account of creation found in the Bible?

Perhaps the biggest disagreement to take place in the Christian church has been over the issue of biblical inerrancy (free from error). Even as I make that statement I recognize that there are those who would define biblical inerrancy differently. Does it mean free from grammatical errors, spelling errors, geographical errors, translation errors or any other error that might be naturally attributed to man before spell check?

As related to our study of creation, could one accept as truth that God created the world but not necessarily in six actual 24-hour days? Does that present a problem with biblical inerrancy? Would that also have a bearing on how one accepts the truth of other biblical passages?

When you hear the phrase, "inspired Word of God", what do you understand that to mean? Are Bible readers intended to search out hidden truth while overlooking factual details?

If one reads the Genesis account of creation, they will discover that the first and second verse of Genesis chapter one actually does leave the option of believing that there could have been a period of time following the creation of a formless heaven and earth before light was created. The actual Hebrew reads:

"In beginning, by Elohiym created the heavens and the earth. Now the earth she was formless and empty and darkness over surface of deep, and spirit of God hovering over surface of the waters" (Gen. 1:1-2).

Thereafter, with the creation of light, the separation of the waters, the production of vegetation, the sun and moon and stars, the creation of animal life and finally the creation of man all take place separately, each in one biblically defined 24-hour day (see Gen. 1).

Thus, one must decide . . . am I going to believe this literally or is this allegory or myth? Is the God of creation able or willing to control biblical content?

It has only been over the past 150 years that biblical inerrancy has even been an issue for the Christian church. However, during that time, and as a direct result of the theory of evolution, everything from Noah's ark to the resurrection of Christ has been called into question. Men have determined that if something does not fit their human experience it is untrustworthy—early church Gnosticism revisited.

I have only presented the issue of disagreement. Hopefully God's Spirit will assist you in determining the answer. I am aware that many try to place themselves in God's role; however, I am also aware that some of their sermonizing, as opposed to correctly handling (cutting straight) the word of truth (see 2 Tim. 2:15), has created much of the problem.

The one who looks for what they believe to be allegory or inconsistencies will find them. The one, who seeks the straight truth, with the guidance of God's Spirit, will find it.

And that is our goal—to strengthen one's faith with the straight truth.

For the Christian, creation cannot be fully understood apart from Christ Jesus, the promised Messiah and Son of God. In the next chapter we will consider how Christ relates to God's creation.

Chapter 5

Christ and Creation

> "He is the image of the invisible God, the firstborn over all creation. For by Him all things were created: things in heaven and on earth, visible and invisible, whether thrones or powers or rulers or authorities; all things were created by Him and for Him. He is before all things, and in Him all things hold together. And He is the head of the body, the church; He is the beginning and the firstborn from among the dead, so that in everything He might have the supremacy. For God was pleased to have all His fullness dwell in Him, and through Him to reconcile to himself all things, whether things on earth or things in heaven, by making peace (restored communion) through His blood, shed on the cross" (Col. 1:15-20).

If Christians truly understood ". . . the mystery of God, namely in whom are hidden all the treasures of wisdom and knowledge" (Col. 2:2-3), those words Paul delivered to the Christians in Colossae would remove any confusion they might have concerning how the universe and life began. Was it in six days or did it take four-and-a-half billion years? Was life created by random chance, requiring multimillions of years, or was it a carefully designed miracle brought about instantaneously for Christ by the Word of God? As a Christian only one answer can exist as to how the world was created by Christ and for Christ. One thing is certain from scripture . . .

Christ Existed with God Before the World Began

> "In the beginning was the Word and the Word was with God, and the Word was God. He was with God in the beginning. Through Him all things were made; without Him nothing was made that has been made. In Him was life, and that life was the light of men. The light shines in the darkness, and the darkness has not understood (overpowered) it" (John 1:1-5).

Many who wonder how the world and life on earth began commit the same fatal error. I say fatal, because, once committed they are doomed to the domain of uncertainty. Some look at creation and reason there is a God. How the world began is of no major concern. They reason that six days or six billion years is all the same to God. Because God exists outside the dimension of time, space and substance He is, therefore, not subject to the restraints of time, space and substance.

The First Question One Should Ask Is Why

. . . but, in understanding why, one should also be given insight into how or, at least, "how not."

God believers fall into two basic categories:

1. Those who believe in God by some definition, and
2. Those who believe in God as revealed in God's Word—the Bible's revelation.

I know that some desire a third category to be "those who believe in God as revealed in some of God's Word or based on other religious writings." Sorry, that just doesn't compute. Once a person decides what part of God's Word or other writings to believe, God then becomes a god of their own word and not God's Word.

God disbelievers fall into only one category—those who dismiss God and instead determine that man is the only source of knowledge and truth. Once again you insist that I have forgotten about the agnostics (those who doubt the existence of God). Sorry, that doesn't compute either. God does not exist for the agnostic as well, because a god without influence in one's life is no god at all.

An Aside . . .

If there is one thing I understand it is agnostics, because at one time or another we all are one. To doubt the existence of God is not unique to the agnostic.

My cousin who died recently and with whom I felt a special love and bond, once proclaimed to me that he was an agnostic. After graduating from college and having attended a Baptist seminary for a time, he now doubted the existence of God. I was of little help to him, because at the time I too experienced my own doubts and was facing my own demons.

Taylor was undoubtedly one of the kindest and most generous men I have ever known, and for years I worried and prayed that he would find peace with God. For many who knew him, he walked the Christian walk with far more compassion than most professing Christians.

Taylor had just turned 50 years old when my wife and I visited him in a Richmond, Virginia, hospital while he waited months for a heart transplant. What brought about the change I do not know, but it was obvious that he was no longer an agnostic. He talked of God and of his favorite hymns. And once again we found common ground in the Lord.

Twelve years after receiving his new heart, I spoke at Taylor's funeral—the funeral service he had written himself. It was a service of joy and celebration, exactly the way he had lived his life. For Taylor literally went to heaven singing his favorite hymn:

> "The church's one foundation is Jesus Christ the Lord. She is His new creation, by Spirit and the Word. From Heav'n He came and sought her, to be His holy bride, and with His own blood He bought her, and for her life He died."

As his fellow professors and friends from the University of Richmond sang that day in the University chapel "she is His new creation by Spirit and the Word", with a smile on my face and tears in my eyes, there was no doubt in my mind that Taylor now rested in the arms of his Creator God.

What Jesus Said

Jesus referred to the book of Genesis 25 times during his recorded ministry. He spoke of how God made both male and female "at the beginning of creation" (Mark 10:6). He spoke of the coming days of affliction as "unequaled from the beginning when God created the world" (Mark 13:19). In Christ's pronouncement against divorce He spoke of God's plan for marriage *"at the beginning"* (see Matt. 19:4-6).

Biblical inerrancy was not an issue with Jesus. He talked of creation, Noah's ark (see Matt. 24:37-39) and Old Testament laws and the prophets (see Luke 16:31) often quoting directly from Scripture. Over and over Christ referred to himself as the fulfillment of Scripture (see John 5:46-47). Never once did

Jesus say or imply that He came to correct Scripture (see John 10:35).

Eternal life and the resurrection of believers both hinge on the literal act of the resurrection of Christ. If Jesus was not raised from the dead by an omnipotent God of life creation, then we are indeed a people without purpose and hope. The Easter miracle, when God spoke and His Son arose from the grave, authenticates God's power over life and creation by His spoken Word.

What Jesus Did

> "Therefore, just as sin entered the world through one man, and death through sin, and in this way death came to all men, because all sinned—for before the law was given, sin was in the world . . . Nevertheless, death reigned from the time of Adam to the time of Moses, even over those who did not sin by breaking a command, as did Adam, who was a pattern of the one to come" (Rom. 5:12-14).

> "For since death came through a man, the resurrection of the dead comes also through a man. For as in Adam all die, so in Christ all will be made alive" (1 Cor. 15:21-22).

A biology professor at Cornell University said, "Let me summarize my views on what modern evolutionary biology tells us loud and clear . . . There are no gods, no purpose and no goal-directed forces of any kind. There is no life after death".

Around the world today young Christians are beheaded, burned to death, thrown in prisons to be tortured and later shot to death. In many cases, we are told, they are given the opportunity to renounce their faith in Jesus Christ and, thereby, be spared from death.

What is the hope that enables these Christians to accept death rather than reject Christ?

What a horrible fraud has been perpetrated on mankind if the sin of Adam has not been redeemed in Jesus Christ? How much of our faith in the sacrificial death of Jesus to atone for the entrance of sin into the world through Adam must be abandoned if Adam never existed at the beginning of created time?

If the sin of Adam is a myth, so too is the redeeming death of Jesus Christ. One act cannot be separated from the other according to the Scriptures.

As always, some will say, "but what about the science?" Soon you will read about the science and, having done so, you will be required to make a decision. It may not cost you your earthly life as these faithful followers in foreign lands. It may not even interrupt your Sunday morning worship attendance, but it will affect your testimony.

Too many of those who claim a relationship to Christ fail to understand how the death and resurrection of Christ is inexorably bound to the creation of man and sin of Adam. If Adam evolved as taught by evolution, then all life evolved. If a portion of the creation account in Genesis is a myth then what is reliable? What plumb line must one use to determine the truth? Why would an all powerful God give us an unreliable testimony?

There are many professors who rely on unproven theories to reject the supernatural activity of God as recorded in Scripture. Their agenda is obvious for all to see. They reject God, God's purpose and God's Son. For many, their challenge of teaching, bordering on obsession, is achieved once they have caused to be compromised the Christian faith of a young believer.

A believer in Christ cannot allow flimsy science to separate them from the love of God in Jesus Christ or the resurrection promised to those who believe.

More Definition . . .

Now don't get anxious and skip over this next part because it is important. Soon we will examine a great body of evidence. The evidence falls into five categories:

1. Scientific evidence—that which can be observed, tested and repeated over and over again.
2. Observed evidence—conclusions that can be drawn from eye witnesses or measured observations of time, space and substance.
3. Philosophical evidence—that which can be logically or rationally accepted as truth and legitimate evidence.
4. Metaphysical evidence—that which falls into the category of the supernatural, outside the realm of natural law.
5. Biblical evidence—that knowledge one gains only from God's Word, the Bible.

Biblical evidence differs from metaphysical evidence in that it is a written source of man's experience with the God of the Bible not simply the supernatural aspirations and/or moral awareness of man.

Evidence Does Not Prove Elohiym

What evidence can do, however, is point one in a right direction and keep him between the lines.

I make no claim of being a scientist; so I am thrilled that so many scientists have done the science stuff for me. Neither do I claim to have the theological expertise of many in the biblical academic world; so I am also grateful for their writings as well. I do have some training in physiology, anatomy, theology and philosophy, though it pales in comparison with many others. What I am is more of a "Joe the plumber", who can understand misleading motives when he hears them, or a Glenn Beck who likes to think for himself . . . perhaps even a Sister Teresa who spends a lot of time listening to God and not the "popes" of the world. And, I read a lot.

Of course, I believe that the God of Jesus Christ is the source of all knowledge and I always pray that He will give me understanding and clarity of thought, especially when I am talking about Him.

"For By Him (Christ) All Things Were Made" (Col. 1:16)

The Bible tells us that all things were created by Christ, for Christ and it is in Christ that all things hold together. Christ was before all things and the fullness of God and the future reconciliation of man to God dwelt in Christ.

Now if you can work that into a formula of creation out of chaos, randomly determined by survival of the fittest, then one of us has missed the "why" of creation according to the Bible and has no clue as to the "how" of creation.

At this point some of you are saying to yourself, "Why should I spend my time reading metaphysical theories about creation when everyone knows that evolution, at least in some form, has been scientifically proven?"

There are many reasons, not the least of which is the thousands of scientists who now say (now that all the evolutionary hype is over) that the theory of evolution needs to be relegated to that place where the theory of a flat earth and bleeding a patient in order to cure a fever now reside. If it were not for the emotional determination of some to keep this theory alive, that is exactly what would have happened by now.

Secondly, maybe it is time that people consider some alternative evidence—yes, I said evidence, not Scripture. I love Scripture and, for me, it makes far more sense than the babblings of the world; however, because there are so many people who have adopted an evolutionary theory of creation, an examination of more recent evidence is what's called for.

Thirdly, Christians need encouragement. Today, more than ever, Christians are confused over what to believe about God and the Bible. For some that confusion has led to a rejection of parts of God's Word and, sadly, apathy over the remainder of God's Word. Even the clergy lacks conviction when it comes to the supernatural miracles in the Bible. If you don't believe me, ask your pastor whether he believes in a Genesis six-day creation or evolution, and then listen to him stutter.

Finally, both of my sons told me that I could not convince them that evolution was untrue. I regret what that says about me as a parent, but it's not too late to try, with God's help.

There is hope, and in my opinion, it does not rest with the clergy but rather the people of God. In America and all around the world people are examining their faith, searching for truth and rediscovering God's Word.

Many are discovering that secular joys and solutions are not the answer to today's problems and anxieties. God is moving in the world and it is truly exciting to watch.

Deliberately Forgetful

The Bible talks about people who are willingly ignorant, also translated deliberately forgetful, (see 2 Peter 3:5). Peter describes these people as scoffers who follow their own evil desires (vs. 3). Ignorance, however, is not the exclusive domain of scoffers. Many in the world as well as in the Christian church find themselves in a state of willing forgetfulness.

Like teenagers who must wear the most popular clothes and flaunt the latest tattoo, many have allowed others to do their thinking for them. And it doesn't take much for that to happen. Our schools and even our churches discourage outside-the-box thinking. Conformity is valued over and above questions and inquiry.

Jesus said "Woe to you experts in the law because you have taken away the key to knowledge. You yourself have not entered (true knowledge) and you have hindered those who were entering" (Luke 11:52). Jesus is saying that some (in this case those who were the religious leaders and experts in the Jewish law) have "locked up" the truth (knowledge). Not only have these leaders locked themselves out, they have hindered others from finding the truth.

We turn now to examine the life and work of Charles Darwin.

"The Learned fool writes his nonsense in better language than the unlearned; but still 'tis nonsense."
—Benjamin Franklin

Chapter 6

Charles Darwin:
The Relationship of Unitarianism, Natural Science and "Origin of the Species"

In an article that appeared in Time Magazine, November, 2006 entitled "God vs. Science", what was to be a knock-down, drag-out debate between two reputable proponents of each cause more closely resembled an exercise in inaccurate stereotypes. It has become increasingly obvious that evolutionists have no idea what people of biblical faith (faith in God's inspired Word) believe and even worse, many professed people of faith have no idea what they themselves believe when it comes to faith and science.

For example, and as presented in this article, it is wrong to assume that only people of biblical faith oppose abortion or embryonic stem cell research. There are many outside the community of biblical faith who oppose both based on respect for life, whether evolved or designed. Neither do people of faith argue with scientific fact or scientific discovery. Evolution, as presented by Darwin and upheld by others, does not fall into either of those categories.

Christian faith is not blind. Christian faith is built upon a real person who lived and died and, according to many eye witnesses, arose from the grave. Scientific fact is determined by experimentation, demonstration and repetition. None of that is present in Darwin's evolution over long periods of time and natural selection. If the goal of an evolutionist is to discredit a six-day biblical faith in creation, they must be reminded that evolution is solely dependent upon faith for its survival as well.

One could only wish that those who purport themselves to be science oriented evolutionists would spend long periods of time understanding the true principles of faith and science, rather than assuming to know what people of biblical faith believe and why they believe it.

Natural Science and the Unitarian Church

To understand Charles Darwin, the naturalist, it is helpful to understand Darwin the Unitarian. Much is made of the fact that Darwin graduated from Cambridge with a theological degree and, subsequent to his early years of study and research in the field of natural science, he rejected his faith and died an agnostic. The implication is that once real science and theology cross paths, only science is left standing.

Natural science is a discipline of observation, labeling and limited conclusions. Natural science does not invent, it does not improve and it does not cure. To imply that observation, no matter in what great detail, is scientific does a disservice to empirical scientific research.

Observation of nature is a respectable field of study; however, it is conceivably only the first phase of a scientific process that must include demonstration and verification in order to draw conclusions based on scientific procedures.

For one to travel to Yellowstone National Park and observe Old Faithful come to life several times each day, one could, with reasonable confidence, predict that it will repeat itself tomorrow. Can one, however, have that same confidence in asserting that Old Faithful performed that same ritual one thousand years ago or one million years ago?

Observation is always an action of the present. It is in the area of past assumptions and conclusions based on present time observations that natural science tends to take a leap of faith apart from the observed evidence.

Charles Darwin, to our knowledge, never referred to himself as a scientist. Darwin traveled extensively from the age of 22 until, taken sick, he returned home at age 27. What Darwin observed amazed him as it might amaze anyone given that same opportunity in 1831. His travels took him to some of the most remote areas of the planet including the Canary Islands, the Falkland Islands, Peru and the Galapagos Islands where he collected many new specimens unknown to the outside world.

It is difficult to suggest what immediate conclusions Darwin drew from his travels except, as he said, that animals and plants were observably "adaptive to their environment," as with Sword-billed hummingbirds with long bills and tongues able to collect nectar and pollinate flowers with long corollas.

Unitarianism

Rather than rejecting his childhood religious beliefs, Darwin's observations reinforced his early religious training.

Unitarianism is a religion of ethics and character. It rejects the inspiration of Scripture as well as Jesus the promised Messiah and Son of God. It is a religion of reason based on one's observation and understanding of nature and life. To the Unitarian, any achievement attained by man is solely dependent upon man's ability to understand his world and adjust his good deeds. According to Darwin's childhood faith, "the natural world is the center of our lives."

Darwin the Scientist

To refer to Darwin as a scientist based on two years of medical school between the ages of 16 to 18 is also a stretch. Darwin was an observer of the present, a naturalist, and his speculations related to animals and plants adaptive to their environment could just have easily been explained as plants and animals having been designed for their environment.

Both explanations are either philosophical or religious assumptions, and do not fall into the category of scientifically verified conclusions. Neither Charles Darwin nor Ralph Waldo Emerson, another 19th century Unitarian well known for his attacks on Christianity as contrary to reason, were present to document the creation of a flower much less the assumed selection process that supposedly brought it to its present form. Nor has any fossil evidence ever suggested that any plant or animal has existed in any form other than its present or extinct form. There are no transitional life-forms in the fossil record.

"On the Origin of Species by Means of Natural Selection, or the Preservation of Favoured Races in the Struggle for Life" (1859)

Having dismissed the Genesis account of creation as allegory (a symbolic narrative) and based on his early years of natural observation, Charles Darwin set about to explain his belief as to how creation might have occurred. In reading his theory of evolutionary progression from lower to higher and/or stronger life-forms in both plants and animals, one should be impressed with Darwin's lack of dogmatism as well as his display of an overall reasoned approach. Read objectively, his work is neither demeaning to people of biblical faith nor does it claim to have all the answers.

Darwin's parents were devout Unitarians and they taught their son to believe that the Genesis story of creation was a myth. Based on that childhood foundation, his theory, though constrained by 19th-century scientific evidence, is quite logically presented as one might expect from Darwin.

Darwin was greatly influenced by George Lyell, recognized as the author of the "Geologic Column" (an attempt to date the age of the earth based on observed earth strata and fossils found between each of its assumed layers). If what Lyell suspected was true, then it was only logical for Darwin to look to the past based on fossils stored in, what was suggested to be, sedimentary layers of an ancient earth.

One Common Ancestor

Based on the likeness of organic beings, their similar embryological appearance as well as what Lyell reported as a succession of lower to higher fossilized life-forms in the earth's strata, Darwin concluded that all plant and animal life began with one common ancestor. Simply stated, Darwin believed that every plant and every animal life-form must have evolved from the first living cell as branches on a single tree or different species from a single seed—"descent with modifications." Scientifically, the beginning of Darwin's theory should have been the end of Darwin's theory, because, unless a tree is grafted, an apple tree will produce only apples and an oak tree will produce only acorns.

The theory of natural selection states that all animal and plant life began as one single life-form that then mutated itself into multiple different and more complex life-forms. The struggle to survive continues when the stronger and more viable life-forms propagate and the weak die even among members of the same species. According to Darwin, mutations, environment and a variety of unknown factors all influence the fight for life. The greatest factor, however, is time—long periods of time allowing for mutations, adaptation, propagation and the change necessary to create a new kind of life-form or species.

According to Darwin's theory and his later writings, species were not created by God but evolved over the years. In addition, evolution was not directed by God but governed by the fortuitous events of natural selection, though those events have never been observed.

The biblical account refers to plants and trees being created fully formed "according to their kinds." Each would in turn produce life-giving seeds to reproduce plants and trees after their kind. The biblical account refers to things of the sea and creatures of the sky being created "according to its kind" (see Gen. 1:11-25).

There is no mistaking the first major conflict between evolution and the biblical account of creation:

1. Either the creation of life began with one single magic cell and somehow overcame the obstacle of

        plants and animals having different DNA and cell structure, or
2. Plants and animals were created as separate kinds from the beginning.

It is important that one not gloss over the first major distinction, as much of what is to follow is built upon the single ancestor theory. Evolutionists will often suggest that there are several alternative theories and that the "design" creation theory, suggesting a divine creator, is the only religious theory and, therefore, not in the field of scientific methodology or empirical science.

The truth is that there are no other alternative theories, and both of the only two possible theories are beliefs, not science.

Evolution is totally dependent upon the one-cell, one-ancestor theory; otherwise, evolutionists must admit that multiple cells containing all the individual atoms, molecules, DNA and RNA chromosomes were spontaneously created like popcorn in a microwave in order to produce each individual plant and animal kind. That alone would destroy the theory of evolution.

In other words, according to evolution, the first life-cell spontaneously created itself out of non-life. Unless this miracle cell also contained the potential DNA programming of all earth's life-forms (plants and animals), then every existing or extinct plant and animal kind must also have its own miracle birth from non-life to life with the entire DNA programming included.

Atoms, molecules, and DNA strands, of which there are multimillions within the simplest single cell, are non-living. It is only their finished design as a functioning cell that qualifies as life.

Question: With today's DNA discoveries how logical is it to believe that frogs and logs came from the same ancestor? How many billions of years must one wait for that miracle to occur? No amount of "acclimatization" or "correlation of growth" (change in one part modifying other parts, which has never been observed) will allow DNA crossover between species, though Darwin suggests it does.

Who Survives?

Darwin asks the question, "How have species or varieties of organisms been perfected?" And the answer is, "They have been converted into good and distinct species."

According to the theory of evolution, organisms must be converted into new organisms and the old organism must die, or the world could not hold all the species. All species must struggle to survive and increase in number in order to propagate and adapt to their environment and, based on the "web of complex relations", become independent of other organisms.

According to Darwin, "natural selection is daily and hourly scrutinizing throughout the world every variation, even the slightest, and rejecting that which is bad, preserving and adding up all that is good . . . at the improvement of each organic being . . . over long periods of time . . . so we can see that only the forms we see today are different from what they use to be."

Natural selection allows one large group to gain advantage over a smaller group through "sexual selection" determined by the dominant life-form; whereby, over long periods of time, the egg or seed is modified to contain the substance of the new and stronger life-form. Darwin suggests that these variations are, once again, due to gradual mutations and environmental influences.

Herein lays the second series of conflicts to which Darwin readily admits:

1. In such a process of new life-forms being constantly formed, where is the fossil record of old "transitional" life-forms? These transitional life-forms should number in the multimillions at least; and yet, none is found. Darwin's response is, "There is imperfection in the fossil record", and at that time in history (1800s) there was indeed a very shallow fossil record. Today's fossil discoveries are hundreds if not thousands of times greater than 19th-century discoveries and still there are no known intermediate life-forms (i.e. a frog with feathers, or a man with a shorter and shorter tail).

For those of you who watch the National Geographic channel you may recall how from time to time they will highlight a recently discovered fossil that indicates to them a transitional life-form. Most recently they highlighted a small dinosaur fossil track indicating what appeared to be feathers, which they assumed allowed this small creature to glide, and presented it as a forerunner to a bird (in this case, a turkey).

Three problems:

      a. To find one or two or forty or fifty supposed transitional life-forms is like saying "I have found

50 specks of sand and conclude that the world is a beach".
b. With every supposed transitioned life-form (say from a lizard to a turkey) there must exist millions of transitional life-forms in between, even if both came from a common progenitor. When you add that to the thousands of distinct plants and animals in the world, man would be constantly discovering transitional fossil remains. Scientists suggest there are approximately 8,000 basic kinds of animals in the world today.
c. Finally, the National Geographic channel failed to inform its listeners that the creature was suspected by many to be a fraud. That is not their agenda.

2. Can modification with one animal produce the complex or even the simpler changes in another? One understands that breeding can produce larger or smaller dogs and cats with more or less hair (microevolution). However, can gradual modifications over long periods of time create completely different species (macroevolution)? Modern genetics say no. Darwin's answer, which of course is the only possible evolutionary answer, is that it must be so because there is no other explanation aside from the supernatural.

3. What about animal instinct? Can instinct be acquired, modified and passed down to later species (those that survive) based only on a process of slow natural selection? For example, how did the salmon acquire its instinct to travel upstream through treacherous waters after mating and before spawning? After spawning the Pacific salmon

die, which indicates more of a propagation instinct than a survival instinct.

> What genetic information taught the Monarch butterfly to lay its eggs on the bottom of a milkweed leaf, the only leaf its larvae will eat; also from which they derive a poisonous toxin for protection from predators during their annual migration from northern climates to southern regions and back north again?
>
> Darwin's answer is that survival in life changes habits; however, both examples above contradict the survival instinct. And for the male black widow spider, its act of propagation costs its life. I recognize that most males will give up a lot for companionship, but where is the struggle for survival? That should have been bred out of the male black widow spider and ended that line of arachnids.
>
> Darwin's answer to which he devotes an entire chapter is that variations of habits are acquired over long periods of time allowing the strongest to live and the weakest to die. Profitable habits evolve in species just as organs (body parts) evolve—there is no other possible explanation for the origin of instinct according to the evolutional model.

Darwin's "Two Great Laws" of Natural Selection

1. Unity of type, whereby an existing or now-extinct life-form is the result of gradual descent from a lesser life-form.

2. The "conditions of existence" have been fulfilled once a type of species has adapted to life over long periods of time and struggle, evolution, stops.

The biblical account bypasses both Darwinian Laws as "the heavens and the earth were completed in their vast array" in six days (see Gen. 2:1-4)—instantaneous creation.

Creationists have no need to explain the lack of intermediate life-forms, as all types were created by a single spoken word, which is exactly what one observes in nature and the fossil record. Survival was not a struggle where countless life-forms must die in order for other life-forms to survive. Instinct was programmed and immediate, and not the result of habitual destructive actions until eventually animals got it right, as if they ever could without the proper programming.

Bees knew how to construct the cells of a beehive essential to their survival. Ants knew how to protect their colony by stealing pupae from the nests of other ants for fighters or food.

Thus far we have not mentioned the codependency of plants and animals; however, our example of the Monarch butterfly provides a perfect opportunity to do so.

Except for the milkweed plant, there would be no Monarch butterfly. Not only does this single plant provide the only food accepted by the larvae for nourishment, it further grants the mature butterfly toxins for protection. It is strange that ecologists today recognize the interdependency of all aspects of the earth's environment; but, at the same time, many fail to see its inconsistency with evolution over long periods of time. They fail to accept the necessity of a fully functioning ecological system in order for any life to survive; unless, of course, the milkweed and the monarch butterfly conspired to evolve on the exact same day.

Concerning intermediate life-forms, Darwin says, "they must have existed" and also their numbers must be "enormous." As to why none have been discovered he, once again, blamed it on the "inconsistency of the geologic record." Later we will discuss the inconsistencies of the geologic record, as, admittedly, Darwin was not exposed to the wealth of fossil discoveries we have today. Perhaps, had he been so, his theory may have evolved differently.

Succession of Organic Beings

Restated, the debate, according to Darwin, is between:

1. Slow and gradual modification through descent and natural progression, or
2. The "common view" (biblical view) of "immutability of species" (existing unaltered and unchanged), except for small changes among "kinds."

In order to sustain a theory based on succession of organic beings, certain absolutes must exist. It does not matter that, according to Darwin, these absolutes are never observed or are inconsistent in the limited observation that is available. Remember readers, a theory need not offer measured observation or repeated experimentation. A theory may sometimes proceed beyond where its assumptions come into question.

Inconsistencies can be explained as anomalies (a deviation from the common rule or form). Lack of evidence can be explained as gaps in the record that will eventually be corrected.

Darwin believed that George Lyell had demonstrated that new life-forms begin to appear throughout the ages as he observed them in the Geologic Column. He further theorized that higher forms of life must change more quickly, seeking to rationalize why transitional life-forms are not available.

He suggests that some of the same fauna appear in different strata, supposedly separated by millions of years, because they are from different regions of the world and evolved at different times.

He further suggests that these inconsistencies might also be the result of a natural catastrophe or glacial period. Darwin does believe, however, that most changes take place simultaneously throughout the world.

Without any intermediary evidence he asserts that, in general, extinct forms differ greatly from existing forms. This assertion is later contradicted by the "Cambrian Explosion" theory.

A Single Birthplace

According to Darwin, "all individuals, both from the same and allied species, have descended from a single parent; and have, therefore, come from a single birthplace . . . and have in the course of time come to inhabit distant points of the globe."

Now we have opened the door to a brand new theory that states at one time in the past all the earth was connected by land allowing for "migration of the species" and adaptation to different regions of the world.

This theory is not inconsistent with the biblical record as far as connected land masses and migration of the species is concerned; however, it does conflict by several hundred million years as to when parts of the world's land mass were connected prior to an earth-changing catastrophe. Later we will discuss the evolutionary theory of Pangaea (migration of land masses from the Greek word meaning "all earth").

Important Early Conclusions to the Debate

1. The debate is not biblical faith vs. science, but rather the biblical account vs. a natural observation of present earth conditions theory.
2. Theory is not proven science, it is only the first step in a scientific methodology that must include demonstration and repetition; whereby, the same theorized results are observed to be repeated over and over again.
3. Whereas Darwin and others would reject "random chance" as the foundation of creation and natural selection and insist that climate, natural environment and unforeseen catastrophic events over long periods of time played a major role in the creation of life and descent of all life-forms—what is that if not random chance?
4. When one's theory must depend on information that may be discovered later in order to fill in the blanks of that theory, it may be time to examine the alternative theory or, at least allow those being educated in taxpayer schools to do so, especially when the blanks that are being filled point away from evolution.
5. At some point one must consider why the vast majority of people believe in a god, and yet deny the Genesis

account of creation, choosing instead to believe in an impersonal god of random chance, chaotic climate change and unnatural selection?

6. Some Judeo-Christians justify their acceptance of the evolutionary theory based on the Scripture "with the Lord a day is as a thousand years and a thousand years are like a day" (2 Peter 3:8) and/or a belief that "God could have created the world any way He desired to do so." The problem with this approach is that it conflicts with the teachings of Christ and the Genesis account in scripture and leaves the door wide open for one to delete any and all Scripture that feels inconsistent with human understanding, which is exactly what has happened.

It is interesting to note that in the passage from 2 Peter 3, as noted above, Peter made reference to the formation of the earth out of water and by water as well as the destruction of the earth by water (the flood).

The reference to a thousand years refers to the Lord's patience in allowing all to come to Him who will come through repentance. While God's patience is long, His Words are sure. "The heavens will disappear with a roar, the elements will be destroyed by fire and the earth and everything in it will be laid bare" (2 Peter 3:10).

Peter then asks the ultimate question: "Since everything will be destroyed in this way, what kind of people ought you to be?" (vs.11). Peter then says, "in keeping with His promise, we are looking forward to a new heaven and a new earth, the home of righteousness" (vs. 13).

7. You may have realized that scientific proof is not the issue in this debate. Evidence, or the lack thereof, logic and, for those who are willing to open their mind, divine enlightenment are the only sources of understanding available.
8. Finally, no matter how many holes logic and evidence may punch into the theory of chance, spontaneous generation and natural selection, the person who is determined to reject the Bible and/or the biblical account of creation will always find a reason to do so. This study is given to encourage the believer to trust the Bible in all things, "and lean not on your own understanding" (Prov. 3:5).

Darwin's Conclusions

1. "Graduation in the perfection of any organ or instinct, which we may consider, either now exist or could have existed, each good of its kind."
2. "All organs and instincts are, in ever so slight a degree, variable."
3. "There is a struggle for existence leading to the preservation of each profitable deviation of structure or instinct." (the good will survive)

According to Darwin, "The truth of these propositions cannot be disputed . . . Natural selection acts solely by accumulating slight, successive, favorable variations, it can produce no great or sudden modification; it can act only by very short and slow steps."

Darwin's theory is logical, it is persuasive and it is consistent with its foundational belief. Evolution over long periods of time, however, does not explain origin of life; it only assumes origin. Origin of life is a far more difficult theory to explain apart from the supernatural as we will discover in the next chapter.

Darwin was confident about the origin of species when he later wrote, "I cannot possibly believe that a false theory would explain so many classes of facts, as I think it certainly does explain. On these grounds I drop my anchor, and believe that the difficulties will slowly disappear."

From time to time, however, that anchor was shaken, as when he wrote in 1873, "The impossibility of conceiving that this grand and wondrous universe, with our conscious selves, arose through chance seems to me the chief argument for the existence of God; but whether this is an argument of real value, I have never been able to decide . . . I am, also, induced to defer to a certain extent to the judgment of the many able men who have fully believed in God; but here again I see how poor an argument this is. The safest conclusion seems to be that the whole subject is beyond the scope of man's intellect; but man can do his duty."

Chapter 7

Natural Law

> "Scientist's religious feelings take the form of rapturous amazement at the harmony of natural law which reveals an intelligence of such superiority that, compared with it, all the systematic thinking and acting of human beings is an utterly insignificant reflection."
>
> (Albert Einstein, 1949)

The Laws of Nature Are Discovered, Not Invented

Where did Natural Law come from? If the universe is the result of a random, never-again observed burst of energy turning a dot of matter into a complex system of heavenly bodies, could this have occurred apart from natural law?

If there were no constant laws in nature, there should be no consistent and trustworthy laws of physics, chemistry or mathematics. If the universe were the result of random chance, why isn't everything haphazard? Why do logic, reason, scientific constancy and perpetual, consistent natural law exist?

While on the one hand evolutionists claim uniformity in nature, they, on the other hand, want to suspend those laws of nature in order to give credence to their non-uniform creation of the universe and life on earth. Some outside force beyond the boundaries of natural, universal law always exists and somehow breaks into our universe at appropriate moments to circumvent the laws of nature and provide for the unnatural (i.e. to create life from non-life or a universe from a dot).

Every attempt to logically explain evolution as a viable alternative to designed creation always includes one or several unnatural acts contradicting the laws of nature. Somehow the laws of gravity, the conservation of energy and biogenesis must be suspended in order that evolution is allowed to make a leap of faith, not science, as confirmed by the laws of nature. Logic and reason must be suspended in order for the unnatural to occur.

The Law of Biogenesis

Based on what man now understands about the complexities of non-living atoms, molecules, DNA threads and RNA-transported directions to protein molecules (nitrogenous organic components required for all life), the Law of Biogenesis states that life may only come from existing life. The theory of evolution concludes that life originated from non-life, and from that first miracle life, all plant and animal life evolved.

The Bible states that life began instantly as an act of God. Plants and animals were fully formed and adaptive to a friendly, life supportive ecosystem. Adam was an adult along with Eve, the woman God brought to him.

Reproduction was originally designed within all life-forms; whereby, life begets life. It just is not possible any other way. Though there are many ways that the theory of evolution can be assailed, the Law of Biogenesis must surely head the list.

Did it take God an entire day to create the solar system or an entire day to create the beasts of the earth? The answer, which is revealed in scripture, is that God spoke and it was so. Why is that significant? The laws of nature, physical chemistry and biology all attest to the fact that neither the universe nor life itself can be created piecemeal. In order for the whole to exist, it must be created as a whole.

Thousands of complex protein molecules are required for even the simplest living cell. Each of these protein molecules contain such a large number of elements, which must be assembled instantly in a required sequence for life to exist. The mathematical chance of even one protein chain being able to assemble itself purely by chance is 1 in 10^{43} powers. When other variables are considered, one author suggests that number is more like 1 in 10^{65} power.

This number only estimates the proper amino acid sequence of one protein molecule; whereas, thousands of protein molecules are required for a single living cell. There are typically over a trillion parts within the average living cell.

Every cell must contain the required basic parts in proper order, structure and sequence in order to be alive. The random chance of a properly arranged sequence of the 20 amino acids capable of producing a functioning set of enzymes (one type of protein) necessary for sustaining cellular life is calculated to be 1 in $10^{40,000}$ power. That's 1 with 40,000 zeros behind it—the greater the number of variables required in a sequence, the lower the mathematical probability of a particular outcome.

Forty cells in a row would stretch across the period at the end of this sentence.

The Magnificent Living Cell

There are those of biblical faith who sharply criticize Charles Darwin as the architect of a monumental fraud. Perhaps it is because I understand how easily one can become enamored with human wisdom and pride, I find myself not so critical.

Based on 21st-century science, Darwin lived under the disadvantage of seeking scientific answers to scientific questions in a scientific vacuum. To Darwin, the magnificent living cell was only a liquid blob. If criticism is due, it might more correctly be directed towards those in the scientific community today who recognize the glaring weaknesses of evolution and spontaneous generation (life from non-life) but choose to remain silent. Their silence has enabled an evolutional theory to continue unquestioned by the general public. Reject God, if you so choose, but at least be honest with the science. To even acknowledge the astronomical improbability of chance creation would be a step in the scientific direction.

The chance formation of only one protein molecule, not to mention the formation of DNA, RNA and life-sustaining enzyme chains, is enough to invalidate evolution, which is solely dependent upon "chance creation" as a theory. Evolutionists will always fall back on the belief that "with a large enough sample any outrageous thing is apt to happen when given enough time and pool of subjects." The problem with that position is that neither the pool of subjects (chemicals) nor eons of time can create a magnificent living cell from non-living atoms and molecules.

1. Who provided the pool?
2. Who provided the chemicals?

3. Who provided the energy?
4. Who said go?

Bear with me a moment as I attempt to simplify the complex process of creating a cell. As already mentioned a typical microscopic cell contains approximately a trillion components. By comparison, the Saturn V Rocket on display at Cape Canaveral, Florida, and stretching the length of a football field is said to have two million components.

In order to create a living cell, certain essential elements must exist.

1. The threadlike DNA gene-coding template (a blueprint) is required to provide the information necessary in order to code an energy-producing protein unit, which in turn provides the energy necessary to code the DNA gene-coding template for its specific function.
2. There are three RNA molecules that now must provide the necessary function of delivering that blueprint to thousands of waiting protein molecules. Messenger RNA molecules receive the code and carry the blueprint. Other protein molecules translate the blueprint (r RNA) and, finally, translator protein molecules translate (t RNA) the blueprint.
3. Protein molecules are necessary to carry out each function mentioned in sequence one and two; and yet, step one and two (DNA and RNA) are first required to code the protein molecules that are needed to perform those intracellular functions.

Here is the dilemma. Protein molecules are required for life to exist; however, a coded protein molecule cannot be created independent of the first protein molecule necessary to provide the energy to code, decode, transport and deliver the coding with the properly sequenced amino acid strands. Life is only possible to be generated and regenerated from existing life; otherwise, there is no life. Functional protein molecules must already exist in order to provide the energy to create other protein molecules required for each step of the coding of all life and all life functions.

Every function supports the whole and every function is dependent on the whole to recreate and sustain the whole. A mutant gene is like a parasitic animal or plant that will eventually destroy the host. Long periods of time and rogue bursts of energy are detrimental to life. They do not enhance life and they certainly do not create life from non-life.

The Law of Conservation of Angular Momentum

What created the universe (the galaxies, planets, stars, moons, etc.)? Evolutionists say it was the result of a "big bang." This theory was presented in 1965 and originally stated that approximately 20 billion years ago all matter in the universe was concentrated into a small dot. Later evolutionists contended that actually there was not even a dot of matter; however, for our discussion it can be a dot or a gaseous no-dot.

At some point an unknown source of energy set the dot (or no-dot) to spinning. All the matter (dust) in the universe swirled and swirled until suddenly it exploded sending matter (galaxies, planets, moons, etc.) swirling into space. Slowly time slowed down these bodies of matter until today we observe them in their present position of motion. The early "Big Bang" theorists suggest that this same cycle will repeat itself every 80-100 billion years. Obviously, there was a very old person who passed on that information.

The Law of Conservation of Angular Momentum states that if a spinning body were to suddenly explode, centrifugal force would cause all its parts to be thrown out in the same spinning direction. If the dot was spinning clockwise, all substance thrown out from the dot would continue spinning clockwise; however, that is not what one observes. Eight of the 91 observed moons are spinning opposite. Some galaxies are spinning backwards.

When this contradiction was noticed, evolutionists attempted to explain this anomaly by suggesting that during the explosion certain bodies hit one another and caused them to spin in the opposite direction. Unfortunately, bodies do not collide in an explosion, but rather get farther and farther apart as they are flung into space according to the Law of Linear Momentum, which states that matter will continue along a straight line if not acted upon—also called the Law of Inertia. In addition, based on angular momentum, the sun should be spinning very rapidly; however, even though the sun contains approximately 99.9 percent of the mass in our solar system, it has only 2 percent of the angular momentum (it spins very slowly).

While scientists no longer recognize this early "big bang" as a viable explanation for the presence of matter, they have nothing to take its place except that which is inconceivable—God. Later we will discuss a new big bang theory that has been described by one scientist as "the fingerprints of God."

While we are on the subject, consider the orderly construction of space—the magnetized core of the earth and other solar bodies, the evolution of tides caused by gravity acting upon the moon and earth's rotation. Consider the orderly placement of the earth to the sun and the moon. Consider the protective atmosphere that preserves life from the sun's radiation. Astronomers have suggested so many stars exist in the universe that every person on earth could own two trillion for themselves; and still scientists today are uncertain as to the composition of a star. The closest star is thought to be 4.3 light years away from the earth or 5.87 trillion miles (one light year) x 4.3 miles from earth. The earth's sun is approximately 93 million miles away from earth.

"He determines the number of the stars and calls them each by name."
(Ps. 147:4)

"When I consider the heavens, the work of your fingers, the moon and the stars, which you have set in place, what is man that you are mindful of him, the son of man that you care for him?" (Ps. 8:3-4)

Scripture teaches that the world was created for man; not that the world might glorify man, but that man might glorify God in the world.

"For we are God's workmanship, created in Christ Jesus to do good works, which God prepared in advance for us to do." (Eph. 2:10)

The earth is both literally and spiritually the center of the universe and man is the centerpiece of creation. In fact, most astronomers today reject the idea that the cosmos has no center, which was the previous cosmological assumption.

Recent findings have contradicted the assumption that the universe was without a center and, in turn, contradicted the old "big bang" theory. This new theory is based on evidence from radio waves or pulsars from supernova (exploding stars) and more recent astrological discoveries suggesting that the cosmos continues to expand.

Of course, the Bible records how God "stretches out the heavens like a tent" (Ps. 104:2) and "created the heavens and stretched them out" (Isa. 42:5).

The New Big Bang

As we will discuss this topic in greater detail in the chapter entitled "The Cosmological Search", allow me to simply identify two recent discoveries: (1) Heat radiation from the "afterglow" of creation and (2) the Great Galaxy Seeds or ripples in the light and heat radiation that indicates to scientists a constantly expanding universe.

Biodiversity

The word biodiversity was originally introduced at an ecological conference in 1986. While not an ordained law of nature, its reality is accepted as essential to the continued existence of life. It is also a chief talking point for ecologists, many of whom believe the earth mysteriously appeared four billion years ago.

Simply stated, biodiversity is the observed variation of life-forms that exist or have ever existed on earth. It is the collective working together of living species with one another and their environment to bring about a healthy and sustained existence on earth. In that sense, it must also be interpreted as "bio-dependency."

Unfortunately, like so many others, those that carry "Save the Planet" banners, by attempting to preserve the diversity of the earth's ecology, miss the forest for the trees. Evolution is the mantra and its theory is recited at every rally. Though it is said that between 5,000 and 10,000 new life-forms are discovered each year, evolutional ecologists continue to teach the extinction of masses of species over the past two to 100 million years.

A mass global change did force the extinction of some life-forms, including most of the dinosaurs; however, biodiversity continues as a natural and necessary phenomenon.

Biodiversity as well as bio-dependency is essential to life. Plants and animals maintain a healthy balance of oxygen and carbon dioxide through photosynthesis and respiration. Pollination is a concert of interdependency between plants and insects with many back-up systems including water and wind. Soil fertilization would continue without the aid of chemicals through decay and waste deposits. Energy is supplied by the sun and the earth's heat and pressure. Flowing rivers provide renewable resources and clean water through natural filtering purification. There are multiple millions of insects, bacteria, fungi and mites all performing essential bio-dependent activities.

Darwin recognized how plants and animals are bound together by what he referred to as a "web of complex relations"; however, Darwin more often acknowledged the predatory aspects of survival that allowed some species to survive in order to serve as prey for other species. "When we reflect on this struggle we may console ourselves with the full belief that the war of nature is not incessant, that no fear is felt, that death is generally prompt, and that the vigorous, the healthy, and the happy survive and multiply." Obviously Darwin never watched "Animal Planet".

The question one must logically ask is "how did one organism, which required interdependence with another kind, evolve separately, separated by one day let alone millions of years?" The problem is not only a one-on-one scenario but rather countless thousands of interdependent life-forms requiring a biologically diverse ecosystem necessary to begin and maintain a biologically diverse planet.

Biodiversity is a God-given reality and a testimony to God's power and His abiding presence in creation. Microscopic molecules give witness to biodiversity as each is dependent upon the millions of its contributing parts to sustain the whole. All creation, from the smallest atom to the largest galactic system, is held together by mutually beneficial distances and functions proclaiming the power of God.

"For since the creation of the world God's invisible qualities—His eternal power and divine nature—have been clearly seen, being understood from what has been made, so that men are without excuse" (Rom. 1:20).

What excuses will the estimated 92 percent* of natural scientists who claim to be either atheists or agnostics give to God when He asks them, "Did you worship the creation or the Creator?"

*This figure represents a significantly higher percentage than among the general scientific community in which assumptions based on universal laws and consistent dependency within the natural order and design are essential to the scientific method of discovery.

Cause and Effect

Can one prove God? Many people of faith have used the premise, which actually began with Aristotle in the 4th-century B.C., that God can be proven based on the Law of Cause and Effect—the first cause being God. What Aristotle actually said was that "motion is the fulfillment of the potential of all that exists." The "essence" of all things exists in "actuality", in a state of potential fulfillment (motion).

The mover (as perhaps an artist) is the agent of fulfillment to the rock (that which is changed into a statue). All things contain the potential for being changed, though not being the agent of change itself.

Because Aristotle viewed all things in motion, it was only natural for him to conclude that something or someone was the first agent of change, hence, the "unmoved mover."

According to Aristotle, science, philosophy, mathematics and religion are outside of matter, and hence, not subject to change. He believed that because they exist outside of matter they must have existed prior to matter. Therefore, because there is motion and there is change, as observed in nature, there must be a principle of first-cause of motion. The first cause is said to reveal the "purpose" of existence.

Later philosophers including Rene Descartes made a distinction between "particular causes" (metaphysics) and "general causes" pertaining to the natural order of things. Descartes stated that causes were only a natural part of the order of things, not the supernatural evidence of a god. To

Descartes, God created the laws of nature, but is not proven by them.

Sir Isaac Newton's Three Basic Laws of Motion and Gravity

1. "Everything exists in a state of rest or uniform motion (as realized in the earth's rotation) unless it is compelled to change that state by force impressed thereon."
2. "The effect of change is proportional to the cause or line of force."
3. "To every action there is always an equal reaction, mutual actions result in mutual reactions."

Sir Isaac Newton (1642-1727) is considered by many to have had the greatest effect on the history of modern science including the later work of Albert Einstein. Newton also denied the metaphysical "first-cause" argument to prove God. He believed that solar bodies were pre-existent and in motion. He suggested that it was gravity that not only brought the apple to earth, but also held the moon and other planetary bodies in orbit.

Newton, considered a radical by the Roman Catholic Church, continued to insist that gravity explained the motion of the planets but not who set the planets in motion. Though Newton adds, "God governs all things and knows all that is or can be done."

The natural law of gravity states that objects with mass attract one another. Gravity keeps the earth, the moon and the planets in their orbit and from colliding.

Today scientists are able to determine the mass of the sun, moon and other solar bodies as well as their distance from the earth based on the law of gravity and their orbital characteristics, thus, one can calculate the distance to the moon and determine that the moon is slowly moving away from the earth and that the sun is giving up mass.

Einstein's Theory of Relativity ($E=mc^2$) is dependent upon the constancy of gravity and the quantity of matter and energy in the universe remaining constant. If that is true, then accidental life is a nonstarter and evolution is only a fantasy.

The Universal Laws of Thermodynamics

Simply stated thermodynamics is the natural process of heat (thermo) being converted to energy (dynamics).

The history of man's understanding of thermodynamics goes back to 1824 as scientists began to recognize how heat expended became heat no longer available—the fire goes out, the heat goes away, a hot cup of tea cools once the source of heat is removed. In both cases heat returns to equilibrium (the same or room temperature as in the above examples).

This observation became known as the Law of Mass Action or Chemical Equilibrium as the flow of a high temperature object to a lower temperature object through a net flow of energy resulted in the temperature of the two objects becoming the same (equilibrium).

States of equilibrium are reached both physically and chemically.

1. Chemical equilibrium provides observable reactions during the stages of mass action which occur between object **A** (heat) and object **B** (the object being acted upon).

2. Physical equilibrium is the natural entropy (decay) process which occurs as an energy source becomes less and less available.

The First Law of Thermodynamics

Energy can neither be created nor destroyed; energy can only be changed. The total energy in the universe remains the same. Energy lost in one process will be gained in another. The Law of Conservation of Energy states that energy is conserved.

Energy exists in many forms including heat, electrical, chemical, light and even caloric. Energy creates work, change and motion as it changes from one form to another.

The Second Law of Thermodynamics or "The Irreversibility of Nature"

Entropy (the natural process of no longer-available energy) in a closed system (our universe) will ultimately result in equilibrium (a state of rest at which no work or change can occur) to increase over time, approaching the maximum value of equilibrium. The natural maximum of equilibrium, at which point no new energy is available to do work, motion and/or change are referred to as "heat death."

The Third Law of Thermodynamics

As temperatures approach absolute zero, when kinetic energy is removed and no longer in use, entropy will approach an absolute zero or constant minimum.

What is interesting for people of biblical faith to observe is that for years evolutionists have claimed that nature, itself, and the passage of time (billions of years) is sufficient to produce the order and complexity of life without an independent (outside of nature) cause. However, as more and more scientists have challenged the theory of evolution based on the laws of nature they have been forced to do the "Texas sidestep"—move in different directions.

Their argument has always been, "you can't prove God" and "we have all this evidence that six-day creation is a myth." Even if that were true, one cannot prove their own theory by debunking another no matter what Socrates said; however, most will agree that creation is one debate where only two alternatives exist: (1) random chance, no matter how that is defined or sought to be explained or (2) by design. One can argue all day long about the nature of the designer, whether the designer is good, bad or indifferent; nevertheless, at the end of the day, only one theory of creation is or can be correct—though, as stated, not scientifically provable.

Could the designer use evolutionary precepts to create time, space and substance, including life? The answer is no, because evolutionary precepts are random and undirected, not the design necessary to begin or sustain natural order.

I realize that many readers will find it difficult to accept the fact that evolution in one form or another (not counting microevolution, large and small dogs) is not real. At one time that was my belief as well. There are, however, both scientific as well as theological reasons to dismiss evolution as a myth.

> 1. Evolution is not science. It does not adhere to the laws of science. Evolution is speculation based on limited observation and desired conclusions. So-called evidence of evolution has a more resonate and reasonable explanation. In order for one to

accept evolution as a "chance" possibility one must be willing to deny all known genetic principles, all known natural laws, all known fossil discoveries and/or lack of discoveries, all known cosmological and biological principles and put their faith in what may be discovered one day that will change all those principles and realities.
2. For the Christian, in order to accept evolution, they must be willing to deny God's Word, to deny the words of Christ, to believe the Bible is sometimes mistaken (especially in something as important as "all things created in Christ") and to worship a sometimes random chance god or a god who waits upon random chance to occur before acting himself.

Thermodynamics Reviewed

The Laws of Thermodynamics state the following:

1. The universe is a closed system. There is no new matter or energy that enters. What we have is all we get.
2. In a closed system, energy used becomes energy no longer available. There is a constant lessening of useful energy. In fact, energy is not lost; it is only changed or transferred into energy that is no longer useful. As energy becomes no longer useful, entropy increases.
3. As entropy increases temperatures approach absolute zero or constant minimum, culminating in "heat death."

What Are the Implications of the Laws of Thermodynamics?

Science today would operate in the dark without confidence in the first and second laws of thermodynamics. They apply to physics, chemistry, biology and geology. Every process ever confirmed through the scientific method was governed by the Laws of Thermodynamics. They are universal laws. All scientific measurements ever made confirm their reality.

In order for life to begin by random chance, the second universal law of thermodynamics must be suspended.

> "All changes are in the direction of increasing entropy, of increasing disorder, of increasing randomness, of running down." (Isaac Asimov, 1973)

Isaac Asimov was a Russian-born and later American writer of science fiction as well as many science textbooks. He was a professor of biochemistry at Boston University and had written or edited more than 500 books. While not opposing religious convictions in others, Asimov was a humanist and believed the Bible represented only Hebrew mythology. He was certainly not a creationist; and yet, he confirms the creationist's position. Asimov asserts that life cannot wait to be arranged for even a second, much less millions or billions of years. Prior to asking the question who or what supplied the energy source, even if that source were available, order will have already tended to disorder, not advance, as evolution requires.

Some evolutionists argue that while the universe is a closed system, the earth is an open system receiving a constant supply of energy from the sun (at least to this point). The problem with that argument is that an influx of heat energy from the sun will produce greater entropy (decay) not enhancement.

In addition, the longer the time span a system is exposed to a solar heat source the greater the entropy. Entropy increases over time, not decreases. Any theoretical molecular structure within such a system will be destroyed not advanced. The mere presence of energy does not generate life nor advance life unless life is already present in its adult or pre-programmed form from the beginning; thereafter, life will advance through reproduction and seed germination. For life to grow in its complexity the proper genetic codes must already be in place or, as one writer said, "contain the proper wiring diagram with high information content."

The Body's Perfectly Designed Electrical System

There are eleven major systems in the human body, which include the respiratory system, the digestive system and the cardiovascular system. All are dependent upon the body's electrical system (the nervous system) to perform and function. As the body's prime communication, command and coordination system, the nervous system is constantly alive with electricity.

If all the nerves from the human body were lined up end to end they would stretch around the world two and one-half times.

The nervous system actually comprises three different systems:

1. The Central Nervous System (CNS) is composed of the brain and its chief nerve called the spinal cord, which extends along the backbone.

2. Forty-three pairs of nerves branch off the CNS and infiltrate every part of the human body forming a network called the Peripheral Nervous System (PNS). The CNS is considered the information center while the PNS is considered the delivery system of information in the form of sensory input for motor output to muscles and glands.
3. The Autonomic Nervous System (ANS), which is partially located in the CNS and shares some of the nerves in the PNS, primarily deals with automatic activities and constant conditions of the body such as blood pressure and heart rate and is independent of the conscious mind.

Nerve cells contain neurons and when these neurons are stimulated by chemicals they produce tiny waves of electricity called impulses. These impulses pass from neuron to neuron with an electrical strength of 0.1 volt and last only one millisecond. Most activity is directed by the brain through the spinal cord; however, certain activity called reflex action is able to bypass the brain. As long as energy is supplied from the blood to the cells this perfectly designed electrical grid will be maintained and life will continue.

Assisted by medical diagnostics, doctors are able to observe the electrical circuitry of the human body in action. They are also able to evaluate anomalies and interruptions as well as sometimes suggest treatment for malfunctions. It remains a mystery, however, as to why or how this system is able to function.

The question an evolutionist must ask is how much faith is necessary to believe that this complex, electrically maintained nervous system randomly constructed itself one gene mutation at a time? Who or what furnished the energy to even begin, much less continue to provide, the electrical energy needed to continue constructing the perfectly designed nervous system? And remember, all those instructions are contained in the single seed of a man and the egg of a woman.

Does Unlimited Time Compromise Natural Law?

Does a process in nature exist that would allow the Theory of Evolution to overcome the second universal law of thermodynamics? Is it possible to overcome the needed complexities of even the simplest cell given the availability of unlimited time? If the pool of all the needed amino acids and other chemical elements were large enough, and the required genes were also present in that pool to provide the components necessary to create the coding for instructions to the RNA transporting threads, could a cell be randomly created from non-life? Based on all that science knows, the answer is a resounding scientific "no", though some scientists insist on holding onto an unscientific "maybe".

The Laws of Thermodynamics state that everything in the universe is in a long-range plan of decay. Man can only hasten that decay, not reverse it. The Bible clearly says that when death entered the world, this world was doomed to die.

James, the brother of Jesus said, "Listen you rich people, weep and wail because of the misery that is coming upon you. Your wealth has rotted, and moths have eaten your clothes. Your gold and silver are corroded. Their corrosion will testify against you and eat your flesh like fire" (Jas. 5:1-3).

Years earlier Jesus had said, "Do not store up for yourselves treasures on earth, where moth and rust destroy, and where thieves break in and steal. But store up for yourselves treasures in heaven, where moth and rust do not destroy, and where thieves do not break in and steal. For where your treasure is, there will your heart be also" (Matt. 6:19-21).

It was God who established the natural laws of the universe. He is not controlled by those laws; He created them. Evolutionists, humanists, rationalists and scientists are given the privilege of studying and living within the boundaries of those laws; however, they have neither the power nor the knowledge to create or discontinue them. Life began with the spoken Word of God and only continues at His pleasure.

Chapter 8

It Takes a Baby

No one doubts the complexities of the human body or what most people refer to as the miracle of birth. Nevertheless, I learned just how miraculous human gestation was from an article by Dr. Randy Guliuzza.

I'm quite sure the information in his article is no secret to the scientific community, in spite of the fact that it contradicts the abortionist's rationale on "free choice" as well as muddies the evolutionary waters.

For years we have been told that the embryo was simply an appendage of the woman who, at her discretion, had the right to cut it off. Evolutionists reinforced that notion as they described the appendage as more fish-like than human.

The miracle of human reproduction actually begins in the male testis through a process called spermatogonia (from the Greek word meaning seed), during which rapidly dividing sperm cells are being limited to only half their normal 46 human chromosomes. It is not until the sperm cell unites with the female egg, which has also undergone the same division of chromosomes in the female ovary, does a new life begin. An additional shuffling of genetic information during division ensures not only a healthy baby, but also a unique individual predetermined at fertilization.

What scientists have also learned is that the baby, or "placenta unit", determines the birth process, not the woman. Once the egg has been fertilized by the male sperm, all future life functions for both the woman and the baby are under the control of the baby. According to Dr. Guliuzza, the baby is not "a passive object being built by the mother's body."

If it were not for the baby/placenta, the mother would expel this foreign object, with its own unique cell structure, as part of the woman's natural immune system. A substance secreted by the baby/placenta suppresses the mother's immune system and allows the baby's acceptance into her body. Equally amazing is the fact that the suppression of the mother's immune system only takes place where the placenta touches the uterus; otherwise, the mother's life would be placed in jeopardy.

This same immunization suppression system protects the mother while allowing her own natural immune system to implant the embryo at the exact depth in the uterus. Without this exact balance the placenta would invade the entire uterus and the mother would die.

Once implanted, a hormone produced by the baby travels in the mother's bloodstream back to her ovary causing it to produce progesterone, which calms the uterus and maintains the pregnancy. All this activity must take place during the initial conception and implantation in the uterus; otherwise, the mother's natural immune system would expel the new life in her body.

Dr. Guliuzza states, "The mother's body is now under the control of a new person." The baby continues to produce other hormones necessary to ensure its survival, which includes "an expansion of the mother's blood volume, an increase in the mother's cardiac output, agents to modulate blood pressure and blood flow to the kidneys along with an increase to the mother's metabolism." In each case the baby's needs are met before the mother's needs.

As the baby begins to reach full term, its production of hormones needed to suppress the production of progesterone that was necessary to prevent early contraction ceases and the baby now produces oxytocin (a uterine muscle stimulant) and labor begins.

As the baby descends within the birth canal, the process triggers a pressure sensor in the mother's brain that increases the oxytocin causing greater contractions. "The placenta now produces the hormone relaxin causing pelvic ligaments and the skin of the birth canal to relax, widen and become more flexible." Weeks earlier, while in the womb, the baby produced the hormones that also allowed the mother to begin producing milk in preparation for its arrival.

Were this study directed against the killing of infants in the womb, the evidence favoring an independent, self-directing baby housed in the mother's body is irrefutable. Human gestation is a duet between a mother and a child, each dependent upon the other for survival.

Can Evolution Reproduce a Baby?

This is a study of evolution and random creation of life. Consider the fact that "survival of the fittest" or survival itself flies in the face of human gestation. Once fertilized, how many mutant genes must be produced in order to allow an adult female to imperil her life with a foreign cell group without her own complex immune system destroying it?

Aside from the miraculous complexity of this reproductive duet performed by mother and child, is it reasonable to believe that an infinite series of accidents (trial and error) produced something that only God could produce? The human embryo produces 15,000 new cells every minute. At any moment the DNA directing that process, the RNA transporting those directions and the billions of cellular particles being formed in perfect order could be interrupted and life immediately terminated. Now evolution must wait for another conception to begin its trial and error reproductive experiment.

The reality is that only a fully created man and woman (male and female) can produce a baby. Life cannot begin from non-life; life must contain all the complex designs needed to reproduce itself.

It's easy to say that anything is possible over millions or billions or even trillions of years, but we are not talking about just anything. We are trying to envision how multi-trillions of complex systems and system ingredients, all of which must exist fully functional and in perfect harmony, could exist by random chance.

All the laws of nature and common sense point to instantaneous, designed creation. It makes far more sense to believe a power outside this universe is responsible and then choose to reject this power as God, than to believe the universe and all sustainable life happened by accident. I prefer to respect this power, maybe even fear this power, and, at the same time, try to know more about this power.

I recall an episode on the old TV Star Trek series in which the crew of the Enterprise discovered a box that was thought to be the source of all galaxies and all life. The problem was that this box was now in the process of destroying what it had once created, so Captain James T. Kirk must find a way to stop it. If I'm not mistaken (and I'm sure someone will let me know if I am) they used reason to convince the box to stop destroying creation—reason that supposedly the box had given its human creation.

The truth is that even a magic box makes more sense than slow, accidental and random evolution to explain life and the universe. My question at the time was—who built the box?

Chapter 9

Fossils and the Geologic Column

According to The Random House College Dictionary a fossil is "any remains, impression or trace of an animal or plant of a former geological age, as a skeleton or a footprint."

Unfortunately, the Random House writers have assumed that animals and plants must wait "geological ages" in order to fossilize. Fossilization occurs whenever organic material is replaced with mineral substance; it does not take "geological ages" for a plant or animal to mineralize.

The Geologic Column is a theory that suggests one might identify the age of the earth according to the sequence of earth strata (layers) and that age is substantiated by the type of fossils primarily found in sediment between each stratum.

Though George Lyell is known for his work in labeling and dating each stratum, ("Principles of Geology", 1830), actually a former Scottish farmer and later naturalist, James Hutton, "Theory of the Earth", proposed a theory in 1795 that said the earth is uniformly layered as a result of a slow, natural process of erosion.

Hutton did acknowledge that in certain locations, more specifically within his homeland of Scotland and the British Isles, these layers were not always uniform, but rather, jagged and curved. He attributed these inconsistencies to the earth's movement, earthquakes and volcanoes—a result of the earth's internal heat expansion.

George Lyell and later Charles Darwin took great interest in Hutton's theory of uniformitarianism, acknowledging him as the first to substantively challenge biblical creation and the biblical age of the earth. Hutton reasoned that a biblical flood and a 6,000-year old earth could not explain the long periods of time he believed necessary to form what he observed as the uniform layers in the earth.

Soon thereafter, George Lyell wrote that he had discovered evidence for what appeared to be a distinct progression of fossilized life-forms from the smallest life-forms he observed in the lower strata to larger and some non-extinct life-forms generally found in the higher strata.

Those observations we will discuss in a moment. For now, however, note that nowhere on earth does the textbook Geologic Column exist in uniform thickness, fossilized distinction or uniform sedimentary layering.

Were it not for Lyell's determination to release science from what he viewed as biblical dogma, his own theory of universalism should have forced him to abandon the Geologic Column for two significant reasons:

1. Even Lyell noted that the earth's strata were far from uniform. Some were deeper and some more shallow. Some were wavy and bled into other stratum.

2. Though he attempted to explain these erratic formations as the result of earthquakes or volcanoes, one must remember, Lyell is suggesting millions upon millions of years to form each stratum, not hundreds or thousands of years. Natural disruptive occurrences like flooding, volcanoes, windstorms, etc. would have created only minor glitches in each rock stratum. Consider the most recent Indonesian tsunami. How significant an effect might one observe in 10 million years in the rock stratum being slowly formed today due to that catastrophe? The answer is . . . none!

Lyell's approximately 600 million years of suggested life history falls into the era that evolutionary geologists now label the Phanerozoic era—originally thought to be the beginning of oxygen-sustained life-forms. Anything prior to the Phanerozoic era is generally referred to as the Pre-Cambrian era and is presently dated back 3.8 to 4.7 billion years. The Proterozoic era (dating back 2.5 billion years) and the Archean era (dating back 3.8 to 4.7 billion years) comprise what is presently taught as the ages of the earth. It seems that once names are ascribed to theoretical ages, they now become reality.

Today what appears in most textbooks as the Geologic Column encompasses the following 12 ages beginning with the age of man and looking to the past.

Quaternary age............................1 million years
Tertiary age................................62 million years
Cretaceous age..........................72 million years
Jurassic age...............................46 million years
Triassic age................................49 million years
Permian age ....50 million years

Pennsylvanian age*....................30 million years
Mississippian age*....................35 million years
Devonian age............................60 million years
Silurian age..............................20 million years
Ordovician age.........................75 million years
Cambrian age.....................45-100 million years

*Not labeled as such in the original Geologic Column

Note that George Lyell only proposed 10 ages. Later dating has suggested 545 to 600 million years and certain ages have been subdivided.

Overstating how pervasive and indoctrinating these evolutional ages of the earth have become in today's world is difficult. They are, it seems, universally accepted and passed on as fact without question. I recently received a flyer from our local alligator farm in St. Augustine informing readers of fossils discovered in Florida dating into the Tertiary era—more specifically the Pliocene age, 10,000 to 2.3 million years ago.

The age of the earth, geological layers and the evolution of life are all tied together in one theoretical package and taught to students as casually as the multiplication table.

However, for those who are willing to consider a second opinion, the fact exists that if any one of those theoretical assumptions is not true, then the remaining assumptions cannot be true as well. There can be no evolution of life without long periods of time. There can be no long periods of time if the geologic time scale is not accurate. There can be no geologic time scale if the fossils discovered between layers are jumbled and not descending in order. There can be no life at all, if life cannot create itself from non-life.

Earliest Evolutional Life

Today some suggest that microscopic life-forms began to appear as far back as 2 ½ to 3 billion years. This new time scale is perhaps evolution's answer to the Cambrian Explosion, during which most fossil evidence based on the Geologic Column indicates an approximate 5- to 20-million-year birth explosion of all present or extinct life-forms (more fully discussed later).

These new and more ancient dates for the beginning of life are based on projections from radiometric dating and fall fully into the realm of "what must be true if evolution is true" and not reliable and/or verifiable science, which we will also discuss later.

Though the theory of evolution appears to be one of natural progression forward through the ages, it is, in fact, a theory built upon looking backward at "what must have been true if the theory is true." For example, if a dinosaur is discovered in the fossil record then something smaller, less complex and less robust must have preceded it. If frogs and birds both exist then there must have been life-forms preceding them and partially resembling them—a "fird" or a "brog."

Every intersection in the road is explained by what must have existed prior to reaching that intersection in order for evolution to arrive at the next intersection. It does not matter whether there is evidence for what must have existed because "it must have existed", and . . . apart from the supernatural, it must have been so.

From the Chart to the Evidence

The appearance of layers or strata in the earth is certainly indisputable; but is it a record of the earth's ages? Before making up your mind, let's examine the visual evidence itself.

Sedimentary rock material is actually those little pieces of eroded earth that make up the earth's surface. The material forms the earth's crust.

Through erosion these earlier sedimentary layers are packed down and will eventually, with the aid of heat produced by the earth and pressure from sediment above, solidify into rock. Rock varies according to the sediments contained in it. Sandstone, shale and limestone are common sediment compositors.

Fossils are, by and large, found in the sediment layers between the suggested rock strata. That is not to say that what is now rock could not have contained fossils at one time; however, those fossils would have been crushed and only small, microscopic tracks might now be recognizable.

Fossils are also found buried in tar pits, and those provide clear and distinguishable life-forms as do fossils discovered buried in tree resin. Igneous (magma from volcanoes) and metamorphic rock (once either igneous or sedimentary rock now morphed into a new kind of rock from pressure and heat) rarely yield fossil tracks or fossilized skeletal remains.

1. Even though some layers are wavy or tilted as opposed to uniformly horizontal and even though these wavy layers sometimes extend thousands of miles, what is referred to as the contact zones (sediment between layers) is most often smooth and consistent. Erosion and weathering, certainly over millions of years, should have created erratic contact zones; and yet, they are generally smooth, suggesting a shorter creation timeframe.

2. Based on present erosion and weathering rates, some scientists project that the present sediment layers of the earth above the continental shelf will be washed (eroded) into the oceans and replaced with new material in 10 million years. Based on that rate, 10 North American continents should have already eroded away in the last 200 million years, thus denying a suspected 600-million-year fossil record.

3. In the Grand Canyon, layers of sediment between strata, projected to be 600 million years in forming, should have long ago washed away. And yet, sedimentary rock remains in spite of the fact that it is clearly exposed to weathering conditions. Eroding underwater sediment layers containing fossils of dinosaurs and reptiles, theoretically dated millions of years ago, show no significant weathering. Sedimentary layers on Mt. Everest, evolutionary dated over 50 million years old, show no significant weathering even at the top. As an answer to these seeming contradictions, some geologists suggest that the Geologic Column, though still millions of years old, was formed as a result of many catastrophic events—thus, they

say, maintaining uniform sediment between layers, which is a new theory to explain a contradiction in the old theory.

Understanding how rock is formed, as observed in the layering of the Grand Canyon, and how impossible it is for sedimentary layers (little pieces of eroded earth) to exist between those strata separated by multi-millions of years of heat pressure and weathering should encourage one to conclude that those strata must have been formed rapidly and more recently. Over millions of years the eroding sedimentary layers should have long since turned to rock or been washed away due to weathering. In other words, unless these rock layers were formed quickly and more recently over a much shorter time, no sedimentary material should remain between them after millions of years of exposure.

An Alternate Theory

As an alternate theory, others suggest that the sedimentary layers between rock strata were laid down rapidly while the underlying layers were still soft during a massive, short-term, catastrophic flood. The mountains we see today were, thereby, being built rapidly.

As waters receded in rocky areas, as witnessed in the Grand Canyon, they formed a stair step effect as a result of repeated short-term flooding.

Some suggest that volcanic action created lava dams that later burst creating each new rock layer, not over long periods of time, but rather over a short catastrophic period allowing for only brief periods of sediment layering.

Various findings around the world have recently indicated that other similar gorges were formed by rushing water as recently as 13,000 years ago and not the original estimate of 180 million years ago when most evolutionists believe that dinosaurs began to appear.

Other places like Monument Valley, Colorado, exist where huge formations stand alone containing sedimentary layers between strata. These formations stand in the midst of vast waste lands, and the most obvious explanation for their existence is that they were carved out by massive flooding, not slow erosion.

Many more examples exist of gorges, canyons and monuments that appear to have been formed by rushing water and are less logically explained by millions and millions of years of slow erosion; and yet, one will not find that alternative explanation in school textbooks.

Remember, evolutionists contend that the bottom layers of each column at one time formed the earth's land surface, and that each layer was subsequently, over millions of years, covered by the next layer. That being said, sediment that formed on the top of each layer would have been either washed away by rain and other natural climate conditions or solidified into rock. There should be no sediment layers at all after millions and millions of years. Sediment layers should have been totally removed; however, this is not what one finds even at the lowest layers.

Assuming that sediment layers could survive over long periods of time, tiny creatures would have buried themselves within those layers to live or feed, thereby mixing up the sediment as was observed after Hurricane Carla in 1961. That, however, is not what is observed. Those sediment layers that do exist are clean and uniform. In addition, one might assume that those layers present in the top strata would have the greatest amount of sedimentary material based on the passing of fewer years; however, they have the least sediment deposits, and as discovered in the Grand Canyon, some have none.

I have mentioned an alternate explanation of how the earth's strata were formed. This explanation points to rapid and catastrophic flooding of the earth. Such flooding would have produced massive dams that would have eventually been broken by greater flooding or seismic activity over and over again. Such a catastrophic flooding would create the placement of fossilized trees in vertical and upside down positions as well as fossilized dinosaurs resting in two or more layers; and that is exactly what has happened and is observed in many locations around the world.

Pause for one moment and consider now what is being suggested based on the evidence of what is actually observed, not what is hoped for to support a natural, as opposed to a supernatural, creation of the earth's surface and life on earth.

1. Why are uniform layers of sediment (non-solidified pieces of shell, limestone, etc.) still present between the layers of exposed rock after multi-millions of years?
2. Why are the suggested geologic layers wavy and not uniform?
3. Why is there not greater consistency in the number of layers around the world? Many places where

layers are observed are missing as many as seven of the expected 10 to 12 strata.
4. How is it possible for fossilized trees (polystrate fossils) to be found vertical and upside down between two or three layers or the protruding neck of a fossilized dinosaur sticking out of one stratum and into another, if those strata were separated by multi-millions of years?

Fossilization occurs when plants and animals are removed from the atmosphere that would otherwise, upon death, cause them to decay. Not only is the presence of polystrate fossils (the same fossil matter existing in two or more strata) a problem for the theory of evolution, it is also evidence of rapid layer formation. It might be possible for a portion of a tree or an animal to be encased in one stratum and fossilize; however, the portion that remained exposed to air over the ensuing "millions of years" would not delay its decaying process long enough for fossilization to occur in the next strata.

What appears far more likely is rapid formation of layers in which trees might indeed be encased in mud between two or more layers and then begin to mineralize.

Nonconformities

It is strange to note that many prefer to hold on to 19th-century observations in lieu of more recent discoveries—discoveries that, in fact, point to completely different conclusions.

Glacier National Park contains a 300-mile series of layers and ridges closely matching the textbook Geologic Column. Evolutionists point to Glacier National Park as a prime example of the layering expected in an orderly column. The problems arose when fossils began to be discovered that yielded a jumbled fossil record. The orderly progression of descending (evolving) life-forms (fossils) is saturated with inconsistencies. The orderly layering was destroying the theory of an orderly progression of evolved life-forms.

Remember, layers are dated according to the fossils discovered in them, and fossils are dated according to the layer in which they are found. A jumbled fossil record indicates that neither the geologic layers nor the fossil distribution are reliable dating methods.

Graptolites, small sea creatures that were dated 400 million years old according to the Geologic Column, are still living today in the South Pacific, unchanged by evolution. Are they now assumed to have stopped evolving over 400 million years ago while everything else continued to evolve? Modern birds are discovered in lower layers than dinosaurs.

One writer wrote that evolution may become "one of the great jokes in the future" as well as "the greatest hoax ever." According to the conclusions offered by those who advocate the earth to be multi-millions to billions of years old and supposedly verified by the fossil record in the Geologic Column, something is amiss in Glacier National Park. What was discovered over 100 years ago is that fossils dated at 150 million years old were found in the pre-Cambrian era dated as much as a billion years ago. No evidence of sliding was discovered between the strata whereby more recently labeled life-forms may have fallen through rock into the Proterozoic era dated a billion or more years earlier.

Other nonconformities continue to be uncovered as modern exploration challenges the textbooks with new discoveries and information. According to Lyell and the Geologic Column, larger life-forms should always be located in higher strata while smaller (earlier) life-forms should only be discovered in the lower strata. This is often the case; however, when so many exceptions to that theory are discovered around the world, one would think that teachers might be allowed to present new and conflicting evidence to their class.

The Geologic Column and its later influence on radiometric dating are dependent on two stratagraphic principles:

1. Biographic—all fossilized life-forms must be consistently found in descendingly more complex forms and,
2. Geographic—all rocks and sediment are consistently dated according to the stratum in which they are found.

Any variation of those two basic principles should cause great concern in the evolutionary community.

Because so many areas around the world have shown a scrambling of life-forms within the various strata, any notion of uniformitarianism and stratagraphic consistency is completely destroyed. These discoveries could not be ignored by naturalists and their quick-thinking explanation is that several short-term world catastrophes must have occurred followed by long periods of quiet time, allowing for a mixture of fossils in strata dating multi-millions of years apart.

This explanation continues to persist in spite of the fact that according to their own theory, rock-embedded fossils are now expected to have migrated from one rock formation to another rock formation separated by millions of years. The evolution of life is assumed to have continued, unabated, during those long periods of quiet time.

PANGAEA . . . is the theory that at one time all continents were joined in two large land masses, and over the course of 200 million years they slowly drifted apart. Visual as well as circumstantial evidence seemed to support this theory. Evolution's theory of species migration was soon tied to the theory of Pangaea to partially explain how similar species exist in continents now separated by water; otherwise, evolution must find another way to explain how life began and how the struggle for life continued simultaneously in regions all around the world.

In addition, the timing of Pangaea is crucial as life-forms must have evolved to an advanced stage of genetic similarity before continents began to separate.

Simply stated some naturalists believed seismic activity led to the gradual separation of two major continents. Oceans and land masses were reformed as these two continents slowly drifted apart.

According to a study by Dr. Arthur Chadwick, Ph.D., many questions have arisen since the theory of "Pangaea" was introduced into textbooks.

1. Movement in water over long periods of time would alter the shape of land mass. What appeared as a

natural fit to the puzzle of separating two major continents is perhaps due more to coincidence rather than slow movement of large land masses. Based on normal water movement, coastlines and land mass would tend to lose its shape were the continents to travel great distances over long periods of time, questioning what some have suggested as a possible visual fit.
2. A more likely explanation provided by later studies of Plate Tectonics is a sudden and massive catastrophic event during which great seismic activity caused rapid plate movement. If occurring on a global scale as a sudden burst of energy, massive flooding and short-term movement would occur.
3. Recent discoveries of deep continental plate roots and the lack of a mechanism whereby long term plate movement might occur have forced some naturalists to revisit the slow-moving, long-term theory.
4. Finally, normal seismic activity experienced today no longer appears sufficient to provide the energy necessary to move large land masses into the currents, and thus, allowing for significant movement.

Darwin's Approach to Migration

Darwin did not envision the movement of continents over large distances. He rather suggested that continents have remained in their present position and were at one time connected by land bridges, perhaps during a glacial period; hence, climate change is primarily responsible for animal and plant migration and slight differences in kind.

Plate Tectonics

Originally the idea of continental drifting was presented by a creationist, Antonio Snider, during the same year Darwin published his work on "Origin of the Species." It wasn't until 50 years later, when scientists began to acknowledge continental drifting, that geologists began to take note of his work.

Recent studies have added a great deal of new information to man's understanding of plates (rigid blocks that divide the earth's surface) and tectonics (movement of the earth).

The earth's surface (the crust) is mostly sedimentary rock in which we find fossils in both the continents and the oceans. Below the crust are layers of solid rock and granite, some of which is exposed to the surface by erosion. What might be called the second layer of the earth is the mantle—solid rock to a depth of 1,800 miles. Below the mantle are the third and fourth layers of the earth—the outer and inner core—made up of mostly iron and molten iron in the inner core.

According to geologists, movement occurs at the edges of the earth's plates in three patterns:

1. Extension (moving apart),
2. Transform faulting (horizontal slippage along a fault line) and
3. Compression (one plate plunges beneath another)—also called subduction.

Based on these three patterns of movement, many geologists have concluded that only massive and catastrophic seismic activity could have caused the separation of what was once a supercontinent.

New oceanic crust is formed by extension, when two plates are spread apart and molten material rises beneath the mantle. Fault lines are created, as in the San Andreas Fault in California, when one plate slides past another. Finally, compression occurs when two plates move towards each other, resulting in one plunging beneath the other— subduction. If this movement is rapid, then the resulting subduction will produce high mountain ranges, as observed in the Himalayas.

A slow and gradual release of seismic energy, as one generally observes today, will not provide the dramatic amount of energy required to move large land masses, raise coastal mountain ranges as well as create deep ocean basins able to contain receding flood waters.

What now appears to be a more scientifically logical explanation of the earth's present distribution of land and oceans is a rapid and catastrophic event as recounted in the Book of Genesis.

Are Creationists Simply Naïve?

Several writers I have read as background material to this study have suggested that those who believe in divine design do not believe in the existence of rock layers. That, of course, would be ridiculous; what believers in design creation have pointed out is that the supposed Geologic Column is irregular, inconsistent and full of anomalies. If people were to imagine perfectly formed or, at least, partially uniform, strata with fossil-bearing consistency (as implied in textbooks) in the Geologic Column, they would be mistaken.

What we do find are many locations with only three or four layers (some with less) and varying degrees of thickness between layers; nowhere do we find a Geologic Column containing all strata and/or strata in uniform thickness with consistent sedimentary layering.

You may wonder how those columns, which only contain a portion of the expected total number of strata, are labeled. Their age as well as their suggested location in the textbook Geologic Column is determined by the dominant type of fossil discovered in them. It is no more complicated than that.

For evolutionists, the age of a fossil is determined by its location in the Geologic Column and the age of a geologic stratum is determined by the dominant fossils discovered in its sedimentary layers. Circular reasoning is not only engaged, it is encouraged.

In fact, there is only .4 percent of the earth's surface where all ten originally labeled strata have been discovered in varying thicknesses. That means that 99.6 percent of the earth's surface has fewer than 10 columns present. Ninety-four percent of the earth's surface has three or more labeled strata missing; 70 percent of the earth's surface areas on land and under the seas have seven or more strata missing.

The above discrepancies and lack of uniformity is definitely a big strike against the entire Geologic Column theory. Multi-millions of years of earth erosion, if it could have been reflected in columns or distinct stratum, would have produced greater universal uniformity. In fact, that is exactly what evolutionists suggested before all the inconsistencies began to be discovered.

When earthquakes, volcanoes, localized flooding or even an ice age takes place, none would be recognizable in the long passing of time necessary for evolution to occur, nor would they allow any noticeable discrepancies on the worldwide number of layers or fossil content. Remember, this suggested period of column formation is the same suggested period of time when all life is said to be evolving, mutation by mutation, in spite of any disruptive activity. Some have suggested a multimillion-year ice age in the past, which is consistent with recent discoveries; however, what happens to the theory of slow, mutation by mutation, evolution during such a time span? Is it possible for evolution to continue uninterrupted? Polar bears might survive but what about hummingbirds?

Slow and gradual, which is the only way evolution is possible, is not reflected in the erratic strata viewed in the United States or around the world. Nor does the erratic mixture of fossils in all layers found in many areas of the world indicate an ancient earth in which plants and animals evolved slowly over millions of years.

The Pre-Cambrian Era (Anything Prior to 600 Million B.C.)

For evolutionists, the pre-Cambrian era is always described as one with a reduced oxygen atmosphere containing a chemically blessed "primordial soup" in which all life began; however, no evidence is available to support such an assumption, only the insistence that it must have been so.

One of the greatest contradictions to evolutional descent centers on the Cambrian era, which contains fossil tracks of life-forms unchanged and present today. Neither can any ancestral life-form be found for the wide variety of anthropoids (resembling humans), brachiopods (crustaceans), mollusks (invertebrates), bryozoans (attached plants), sponges (plant like marine life), annelids (worms) or chordates (vertebrates) living today and also discovered in an era supposedly dated up to 600 million years ago. According to natural selection, such ancient life-forms should not exist today and certainly not in their present form. It would seem more to indicate an immediate beginning to life rather than millions of years of descent from a common ancestor. These discoveries have forced evolutionists to devise a new theory that states that most life-forms are now said to have evolved rapidly over a period of 10 or 20 million years during the Cambrian era (560 millions B.C.).

Plain Talk

1. Devout evolutionists, while admitting to the charge of circular reasoning (in relation to stratum-dating according to the fossils and fossil-dating according to the stratum in which they are found) insist that such reasoning is necessary in order to correctly identify the age of both. In addition, they employ the same reasoning when dating irregular strata in depth and the lack of uniformity in the fossil record and sedimentary layering.

   For example, if one were to accurately measure the vertical (top to bottom) depth of strata discovered around the world and label them as accurately as possible according to the dominant fossils discovered in them, a column would exist as tall as 100 miles (based on the tallest of each geologically dated stratum) or as shallow as one mile, according to the smallest of each geologically dated stratum.

2. Naturalists are constantly discovering life-forms thought to be extinct millions of years ago. The horseshoe crab is a prime example. This arthropod was dated in the Silurian era (424 million years ago) and was assumed to have become extinct 50 million years ago; however it survives today. The mystery is not only that a life-form might survive over multimillions of years, but also that it would survive in its present form without any evolutionary changes. Consider the fact that a single cell in the horseshoe crab is more complex than any super-computer. This is not what one might expect in a creature so closely dated to the dawn of accidental life.

3. Evolutionists insist that some scrambling of the fossil record within the earth's strata is to be expected. Seismic activity would cause some seepage from cracks between layers. The problem with that theory is that earthquakes and volcanoes occurring over millions and millions of years would produce an equalizing effect. In other words, unless these earthquakes lasted several million years in duration, thereby killing all surrounding life-forms, they would have little effect on the layering of strata and preservation of an accurate fossil record over millions of years. However, a world catastrophe as revealed within the Genesis record would certainly produce such a scrambling.

4. A catastrophic flood of short duration (say, one year) would produce a scrambling effect as well as the effect generally noted during which smaller or lower dwelling life-forms settled at lower levels and larger life-forms settled at higher levels. Both larger plant life as well as animal life would be expected to survive the massive flooding longer than smaller species and thus be generally located at higher elevations.

   Remember it is not only flood waters that destroyed life but also the accompanying volcanoes and earthquakes that buried or swallowed up life. Where possible, some larger animals, including humans, would have delayed their destruction for a time by climbing to higher elevations.

5. Overall it is more logical to believe, based on what is observed today, that erosion over long periods of time could not have created earth's present topography or what is observed in the fossil record. Even intermittent short-term catastrophes accompanied by long quiet periods cannot account for the dramatic layering observed in the Grand Canyon. Still the outdated formula of uniformitarianism as a gradual process of erosion responsible for earth's geologic formations persists in textbooks. It is the one theory evolutionists cannot afford to abandon.

6. As mentioned from the beginning, the observed data is there for all to see. It remains a matter of choice as to whether one will follow the evidence to its logical conclusion. One must constantly be reminded that

evolution and the Geologic Column are theories bordering on obsessions and not scientific fact. Sure, there are layers and there are smaller and larger life-forms, but the issue is: did all those life-forms evolve from a common ancestor over millions of years or were they instantly created independent of each kind by a designer God who created them and the ecosystem in which they were able to reproduce and survive. There is no such thing as a little evolution. For the theory of evolution to be true, it is all or nothing.

7. Evolution will persist as a theory of creation in spite of contradictions, anomalies and more recent and future evidence. The true issue is more devious than simply teaching children the difference between theory and fact. In this case, society has conceded theory to be fact and it may take another 150 years to turn that around . . . if God allows the world another 150 years.

Christians are reluctant to suggest they believe in a six-day creation because society is so convinced that it is a myth. Teachers who know better are forbidden to teach any alternate evidence or obvious inconsistencies in the theory of evolution; therefore, theory has become fact by the absence of dissent.

No, God did not leave creation up to chance. God did not allow intermediate life-forms to die in order that the stronger might survive. God did not hurl the stars into space and wait for some random occurrence in order for a cognitive man to eventually evolve from a worm or whatever.

Someone has accurately stated that man once judged life according to the Bible; whereas, now man judges the Bible according to man's human assumptions about life.

Those who do not believe that a god exists do not bother themselves with the absence of facts or evidence. Contradictions and anomalies, missing links and the lack of intermediate fossil evidence are only inconveniences that do not alter the bigger picture. Whenever the possibility of a divine being is dismissed out of hand, then only their inconsistent theory is left standing.

Transitional vs. Intermediate Life-Forms

According to evolutionists, and simply stated, a transitional life-form is one that exhibits a significant number of similar traits to its closest ancestor. A transitional organism is morphologically similar in form or structure to its derived (from which it evolved) relative. A wolf, a fox and domesticated dogs would be considered transitional by most evolutionists.

According to evolutionists, an intermediate life-form is one that displays a large number of unique traits that are not connected to its derived relative. For example, a creature that displays traits of a fish and that of a tetrapod (a four-legged creature) would be considered an intermediate life-form, as would a non-human ape creature to a human (a missing link).

Six-day creationists, for the most part, tend to link both terms (transitional and intermediate life-forms) together. They argue that the fossil record fails to demonstrate a line of derivation between any one ancestor to another that would even partially indicate a substantive change in form or structure. There is no evidence of invertebrates (without a backbone) becoming vertebrates even though evolution claims it took place multimillions of years ago. Where are the transitional or intermediate stages necessary for this transformation to occur?

There are no fin-to-feet creatures in the fossil record. Both land and sea mammals abruptly appear in the fossil record with no transitional ancestors.

To simply imply that a carp might be the ancestor of a duck or a cow is baseless speculation without intermediate evidence. And why are both still present if one slowly evolved into the other? And if they both had the same common ancestor, where are the fossilized lines of descent from that ancestor?

Evolution's Response

One should not expect to find all the evidence that proves evolutionary descent, as the fossil record is often found lacking in such life-forms. Perhaps the intermediate stages were very short and the completed life-form, thereafter, made no significant changes in form or structure. It must be so.

Next we will discuss the evolutional theory that destroys evolution as a theory.

Chapter 10

The Cambrian Explosion

According to Lyell's textbook Geologic Column the Cambrian era lasted approximately 45 million years; others suggest more in the vicinity of 10-20 million years. Some suggest as few as a five-million year span, though it seems ridiculous to suggest how even a devout evolutionist could be that restrictive. Several reasons exist for why a clarification was needed to explain what archeologists are now discovering, namely:

1. The fossil record is consistently revealing that life-forms are mixed throughout the earth's suggested 10-12 strata and suggested 600-million year history.
2. Creatures supposedly evolved 500 million years ago either still exist in their same physical form or have become extinct, as opposed to descending into new life-forms.
3. No evidence exists of any transition taking place among species.
4. The Cambrian stratum contains virtually every phyla (all major subdivisions of the animal kingdom) known to man, all existing in this one lower geologic layer. Rock below Cambrian has virtually no fossilized specimens, though some evolutionists insist they have found microscopic evidence of life prior to the

Cambrian stratum in fossil tracks or rock imprints; six-day creationists do not disagree with those possibilities.

5. As one continues up the evolutionary Geologic Column, there is a dramatic decrease in all fossilized species until the higher layers reveal the unthinkable for evolutionists—98 percent of everything that has ever lived is extinct without any evidence of evolution. This, of course, is consistent with a catastrophic flood but not slow, gradual transition of species.

It was Harvard professor, Stephen Gould, a staunch evolutionist, who first addressed these discrepancies by announcing what he explained as "punctuated equilibrium." In his explanation he was forced to conclude that somehow "life-forms appeared all at once and fully formed." He further stated that animals must have evolved quickly over shorter periods of time (explosion); thus, he attempted to explain the lack of transitional life-forms throughout the geologic record. Of course he is unable to explain how this might be possible.

Gould, as one of evolution's greatest gurus, became the champion of what many believe completely destroys any confidence one should now have in evolved life.

There are no transitional life-forms in the fossil record.

Many fossilized life-forms supposedly existing 60-500 million years ago still exist in their same form today. It should not go unnoticed that natural and biological science is destroying Darwin's theory of evolution, not faith.

1. Evolution is impossible if genetic crossover is impossible.

2. Evolution is impossible if a common ancestor for all plants and animals is impossible.
3. Evolution is impossible if gradualism (slow evolutionary changes over long periods of time) is impossible and unsubstantiated in the Geologic Column.
4. Evolution is impossible if molecular gaps exist between amphibians, fish, reptiles and mammals—not to mention molecular gaps between animals and plants. There can be no crossover between kinds.
5. And finally, evolution cannot exist if the "Cambrian Explosion" is real. According to their own tenets, 5, 20, or 100 million years is not long enough for gradual, chaotic mutations to create all the species we see today or have discovered extinct in the fossil record. The Cambrian stratum was formed too quickly according to Lyell's own theory of geologic layering for evolution to occur.

Naturalists, geologists and evolutionists have all responded to the suggestion that all life-forms exploded in one short time era with their own statements of dismay. It was Darwin, himself, who said that such a discovery would be a "serious difficulty" and a "fatal objection to the belief in transmutation of species . . . and to that I can give no satisfactory answer."

According to Stephen Gould, "The old rationale about undiscovered sequences of smoothly transitional forms will no longer wash." "Science Progress" responded that if evolution is assumed and special creation at the same time rejected that "the absence of any record whatsoever of a single member of any of the phyla in the pre-Cambrian rocks remains as inexplicable on orthodox grounds as it was to Darwin."

Take Down that Family Tree

The imaginary "Tree of Descent" where all plant and animal life slowly evolve from one creature into another more advanced creature has been axed at its roots. Visual similarities based on morphological (form or structure) appearance have all been discounted as lacking in supporting molecular evidence. Pictures and charts are not scientific evidence. Even between those creatures once thought to be close relatives, genetic differences have precluded any crossover. If crossover is denied on the molecular level between creatures morphologically similar, how could one possibly entertain the notion of DNA crossover between the vast amounts of life-forms in the world today?

And take that one step further. If indeed the mass of fossils gathered from all suggested layers of the earth (be they the unlikely result of slow erosion or a catastrophic flood) demonstrates an explosion of all life during one brief period of time (as it clearly does), then evolution needs to be removed from our school textbooks. Leave the teaching of creation to the church where it belongs and bring science back to the classroom.

2. Evolution is impossible if a common ancestor for all plants and animals is impossible.
3. Evolution is impossible if gradualism (slow evolutionary changes over long periods of time) is impossible and unsubstantiated in the Geologic Column.
4. Evolution is impossible if molecular gaps exist between amphibians, fish, reptiles and mammals—not to mention molecular gaps between animals and plants. There can be no crossover between kinds.
5. And finally, evolution cannot exist if the "Cambrian Explosion" is real. According to their own tenets, 5, 20, or 100 million years is not long enough for gradual, chaotic mutations to create all the species we see today or have discovered extinct in the fossil record. The Cambrian stratum was formed too quickly according to Lyell's own theory of geologic layering for evolution to occur.

Naturalists, geologists and evolutionists have all responded to the suggestion that all life-forms exploded in one short time era with their own statements of dismay. It was Darwin, himself, who said that such a discovery would be a "serious difficulty" and a "fatal objection to the belief in transmutation of species . . . and to that I can give no satisfactory answer."

According to Stephen Gould, "The old rationale about undiscovered sequences of smoothly transitional forms will no longer wash." "Science Progress" responded that if evolution is assumed and special creation at the same time rejected that "the absence of any record whatsoever of a single member of any of the phyla in the pre-Cambrian rocks remains as inexplicable on orthodox grounds as it was to Darwin."

Take Down that Family Tree

The imaginary "Tree of Descent" where all plant and animal life slowly evolve from one creature into another more advanced creature has been axed at its roots. Visual similarities based on morphological (form or structure) appearance have all been discounted as lacking in supporting molecular evidence. Pictures and charts are not scientific evidence. Even between those creatures once thought to be close relatives, genetic differences have precluded any crossover. If crossover is denied on the molecular level between creatures morphologically similar, how could one possibly entertain the notion of DNA crossover between the vast amounts of life-forms in the world today?

And take that one step further. If indeed the mass of fossils gathered from all suggested layers of the earth (be they the unlikely result of slow erosion or a catastrophic flood) demonstrates an explosion of all life during one brief period of time (as it clearly does), then evolution needs to be removed from our school textbooks. Leave the teaching of creation to the church where it belongs and bring science back to the classroom.

Chapter 11

The Living Cell: Atoms, Molecules and DNA

As he entered his favorite English pub, Francis H.C. Crick boldly announced, "We have discovered the secret of life." While Crick's discovery, along with colleague James D. Watson, was indeed monumental, it was not, the "secret of life."

What they did discover in 1953, for which they received the Nobel Prize, was the double twisted helix of DNA (deoxyribonucleic acid). Their discovery soon became the template for understanding the structure of molecules that contain the genetic coding of all life.

Evolutionists claimed it as the breakthrough for which they had been waiting to prove the mechanism of natural selection from a common ancestor. Some called it the "language of evolution."

Later years of research have yielded greater knowledge as scientists are now more able to understand and interpret (decode) the molecular genes responsible for human growth and individual characteristics. What is now known is that three essential molecular building blocks exist that must work in concert in order to design cellular structure in an orderly process of growth and eventual cellular and species reproduction.

1. DNA is the thread-like molecule containing the genes (the code) located in all cells.
2. RNA (ribonucleic acid) is the messenger molecules that transport in three distinct phases the coding template to the protein molecules—the energy source.
3. Protein linear polymers (strands of molecules built from 20 standard amino acids in exact sequence) produce the energy necessary to maintain cell structure, growth and reproduction.

After years of continued research in which Crick sought to discover the real "secret of life", he is reported to have acknowledged, "An honest man, armed with all the knowledge available to us now, could only state that in some sense, the origin of life appears at the moment to be almost a miracle, so many are the conditions which would have had to be satisfied to get it going." Nevertheless, Crick also asserted, "people like me get along perfectly well with no religious views."

Atoms, Molecules and Cells

Molecules and the atoms that make up molecules are not alive. DNA, RNA and protein molecules are not alive. Life is created at the cellular level as non-living atoms and molecules perform their designed function in concert with one another and their environment. A molecule may contain as few as two atoms or as many as millions. A cell, of which anywhere from 60 to 100 trillion make up the human body, may contain a trillion or more molecular particles.

Just as the human body has organs (a heart, lungs, kidneys, etc.) to perform specific tasks, each individual cell has organelles that specialize in life-giving functions within each cell. The nucleus (one cellular organelle) contains the DNA that directs the production of energy-producing protein molecules necessary for growth and replication of cells.

Life is not possible in its cellular form unless all the trillions of non-living parts are designed and functioning together. Even a single amino acid peptide needed to provide the energy to reproduce a functional cell could never accidentally form itself in proper sequence from a pool of all the standard amino acid molecules, no matter how long a period of time might elapse or how big that acid pool might be. In spite of that, Francis Crick held to the belief that the mechanism for DNA coding "may be partly the result of historical accident."

Correct Observation—Wrong Conclusion

On winning the Nobel Peace Prize in 1962, Francis Crick presented his latest book, "What Mad Pursuit", in which he documents his religious journey and presents his atheistic, non-belief pilgrimage.

Crick's parents were members of an English Congregational church and attended regularly. At the age of 12 Crick recalls becoming skeptical over the existence of God. Crick later attributes his loss of faith to a growing attachment to science, evolution and the suggested old age theory of the earth. To him it seemed that they all contradicted the biblical account; therefore, he concluded it was "impossible for any balanced intellect to believe in the literal truth of every part of the Bible in the way that fundamentalists do."

Crick ultimately determined to seek the answers to the mysteries of life and human consciousness in the field of molecular biology and not religion. He decided to rely on science to prove or disprove the Bible. Following his discovery of DNA coding through the double-twisted helix DNA strand, Crick spent much of the remainder of his life searching for the answer to one of those mysteries—the origin of human consciousness.

Finally, after many fruitless years of research and great financial investment, he realized that consciousness could not be the result of DNA mutations or any other biological cause.

While continuing his search, in 1971 Crick attended an extraterrestrial intelligence seminar in Soviet Armenia, after which he formulated his theory of "panspermia" (alien beings must have planted the seeds of life throughout the universe billions of years ago)—though he admitted his theory of "directed panspermia" was "highly speculative."

Recognizing the huge hurdles necessary to create life from non-living matter, he later modified his theory to "life itself", in which aliens were said to actually ship certain whole life-forms to the earth millions of years ago. While Crick's alien theory is a little off-the-wall, what it does show is that one of the world's greatest DNA experts recognizes how futile the theory of evolutionary chance creation is—life from non-life or DNA-generated human consciousness.

Some have noted that Crick was a habitual user of LSD during those years of his life. Others have questioned why aliens would plant or ship atmospherically adaptive life-forms to earth that would require millions or billions of years to evolve. Why not just colonize and inhabit the earth? Thus far no alien fossils have been discovered.

Francis Crick died in 2003 at the age of 88, having never discovered "the secret of life." Though it should have been more obvious to him than to others not as familiar with the complexities of life, it is said that he remained an atheist until his death. True to his rejection of designed creation Crick wrote, "What gives biological research its flavor, is the long continual operation of natural selection . . . if this was produced by chance alone without the aid of natural selection it would be regarded as infinitely impossible." Which begs the question, how would he explain the Cambrian Explosion?

Just Consider

If the information contained in one person's DNA were typed out into books, it would fill the Grand Canyon 40 times over. If it were stretched out and tied together it would stretch 10-20 billion miles and complete up to forty thousand round trips from the earth to the moon; and yet, that same DNA would fit in two tablespoons.

Mathematicians have calculated that the probability of one single DNA strand arranging itself by chance is 10 to the $119,000^{th}$ power.

Life Codes

The sides of the DNA twisted ladder are formed by alternating deoxyribose and phosphate molecules and the rungs of the ladder consist of four nitrogen bases—adenine, thymine, cytosine and guanine. This might more easily be understood if one will compare a DNA strand to our own 26-letter English alphabet and how those 26 letters are used to communicate words, thoughts and instructions.

The unique genetic makeup of any and all life-forms is determined by the sequence of these four DNA letters that are identically sequenced in all the thousands of DNA strands in a single cell and the trillions of DNA strands in an entire body.

Darwin once said, "If it could be demonstrated that any complex organs existed which could not possibly have been formed by numerous, successive, slight modifications, my theory would absolutely break down." Now we know that all organs and systems fit that description, including the millions of molecular subsystems in every living cell, and Darwin's theory does indeed "break down."

Within each human cell there are 3,000 million pairs of the genetic alphabet (A, T, C & G) and the adult human body contains trillions of cells while making millions of new cells every minute. This irreducible complexity is something Darwin could have never imagined in 1859.

The Living Cell

In one of his essays, C.S. Lewis states that "every age, within limits, gets the science it deserves." And so it is, man's desire to be his own god has brought about the acceptance of a myth that allows him to reach that aspiration for a time.

Too many Christians make the mistake of believing that the biblical record should be defended apart from the supernatural activity of the God of creation. Why would a worldly man be more willing to believe in the miracle of redemption and eternal life than he would the miracle of creation? Is it not a mistake, on the one hand, to explain away all the biblical miracles that appear to us too unbelievable, and on the other hand, invite someone to experience redemption at the foot of the cross and eternal life after death?

The world, with all its ungodliness, is no more interested in a small god than in the great God of all creation. Natural man is sinful and self-centered. Natural man delights when the God of creation is explained away by the very ones who claim a relationship with Him. God forbid the Christian church from destroying the inspired Scripture in an attempt to make them more palatable to the natural man.

As stated earlier, biblical creationism and the theory of evolution do not fall into the category of verifiable scientific experimentation. Creation, whether over long periods of time by spontaneous generation or in six days by the spoken word of God, will only happen once. It cannot be observed, scientifically verified or repeated. Many have tried and all have failed. Therefore, a belief in evolution also belongs in the category of philosophy or religion. What the proponents of evolution have done is to proclaim themselves scientists and those that believe in a six-day creation as "people of faith." Mind you, I don't take exception to being called a person of faith; however, having studied evolution, it would take a lot more faith to accept the evolutionary myth than the God of Creation. The truth is that both fall into the category of faith ("confidence in things unseen," Heb. 11:1).

Francis Crick speaks constantly of the "unsolvable mysteries of nature," as if nature were a godlike entity that weaved the biological complexities of life devoid of personal involvement or emotion. Crick warns his readers, "Biologists must constantly keep in mind that what they see is not designed but rather evolved." In other words, don't be influenced by the reality of what you are actually observing.

Don't be fooled when the evolutionist says he does not rule out the existence of a god. His entire theory excludes a biblical

God, and only a wimpy Christian would ever say "perhaps God was involved in the evolutionary process". The Judeo-Christian faith is based on an all-powerful, omnipotent God of all creation. Anything less is unworthy of our faith and allegiance. Are Christians to believe after 3,500 years of accepting the inspired, written biblical account of creation, that evolutional theists finally explained the truth about God's six-day creation?

By faith, evolutionists believe life began in a chemically endowed primordial soup. Subsequently, over long periods of time inorganic atoms and molecules randomly turned into a living cell, and eventually all plant and animal life we see today evolved through a process of natural selection. One would expect to now read "and they lived happily ever after."

Scientific Evidence

Simply stated life depends on the energy and proper arrangement of three classes of molecules to provide the information necessary for a cell to feed, protect and propagate itself: DNA molecules that make up the master plan of a cell, RNA molecules that transport the DNA code information of that master plan in three distinct stages and protein molecules that are engaged and required in every step of the process to create the energy to code the cell. All three complete systems are required and all must be created at the same moment with all their parts functioning properly for life to exist. In other words, some components cannot be created and sit wiggling on the sidelines waiting for the rest of the components to be assembled. They must be created simultaneously as each component is critically involved in its own existence and replication.

1. We are told that the creation of the first mystery protein molecule originated in a pool of all the required subparticles that included the 20 standard amino acids necessary to produce an energy-providing protein molecule. This nonfunctioning, inorganic molecule must wait for the correct 20 amino acids to line up simultaneously in correct sequence in order for one protein polymer chain to appear—and that is only the first of hundreds of peptides required to produce one living cell. Multiple proteins are required to form a cell, which includes a cell wall, channels, receptors and all internal organelles needed for reproduction.

   One may assume that such was possible over billions of years; however, it is not as easy as beginning with one standard amino acid and waiting for numbers two and three, etc. to correctly align themselves over time.

   Unless all 20 amino acids align instantaneously in the correctly coded order to create this one still non-living polymer chain, it will not be capable of performing its now unknown function as it waits for other potential protein molecules to also receive their DNA instructions and RNA translation to randomly form.

   DNA's key function is to provide the information necessary to build protein as enzymes or the structural molecules of the body's major organs, muscles and/or other physical units. Assuming a protein molecule has been successful in correctly aligning all required amino acids, we must now hope that the thousands of other potential protein molecules needed to provide information, energy and structure to one cell are also instantaneously successful; otherwise, none will survive.

The greater problem than even that impossibility is who or what is giving direction to even begin this task. Thus far no DNA or RNA protein molecules have been assembled, and their spontaneous formation is far more daunting when one considers that it is the yet-to-be-created energy-providing protein molecules, containing the correct amino acid sequence, that allows the DNA and RNA molecules to form and function.

At least 75 different protein molecules are required to harvest the energy necessary to begin and continue the DNA and RNA coding process—a process in which the 75 energy-producing protein molecules are also coded and created. The odds of even a smaller than average potential energy-producing protein molecule forming by chance is estimated to be 1 in 20^{100} and the odds of all 75 proteins of the same size forming by chance are 1 in 20^{7500}.

Amino acid molecules are the building blocks of protein. When combined in a sequence or polymer chain they are called peptides or polypeptides (larger chains) and/or proteins. The process of creating a protein chain of amino acid molecules (which determines the function of that protein) is called translation. There are between 20 and 23 standard amino acids depending upon how a particular amino acid molecule is classified. In addition, there are also "nonstandard" amino acids present in living organisms, though not in proteins or, having once been in protein, later chemically modified through a process called "post-translation."

Biologists agree that 20 standard amino acids exist in all proteins. The order and function of a protein molecule is genetically determined according to its individual sequence. Without a genetic coding there is no order and, therefore, there is no function.

If the earth was a mass of amino acids it would contain an estimated 3.27×10^{49} amino acid molecules. There are 1.27×10^{130} possible combinations available in forming a smaller than average 100-amino-acid residue or polypeptide chain. It would take an average of 2.67×10^{65} attempts per second in order to approach the proper and metaphysically theoretical assembly of one protein molecule.

2. The DNA molecule is far more complex. Even the simplest cell has 470 DNA genes, which are responsible for coding protein molecules containing about 347 amino acid strands. The probability of coding only one DNA-instructed protein molecule is 10 to the 451st power. If you desire to get the full impact of this figure write 451 zeros on the blackboard and that will potentially code one out of hundreds of required DNA-instructed and functional protein molecules in one cell.

3. Amino acids, the building blocks of protein, require carbon atoms to exist. If the earth were 100 percent pure carbon atoms, it would contain 10 to the 50th power carbon atoms. That means the earth would only

contain approximately one tenth of the needed carbon atoms required to potentially create one protein molecule.

4. Amino acids are genetically coded in various sequences to perform a vast variety of tasks. This genetic coding determines all biological functions including reproduction. In order to produce a cell with a self-replicating system (reproduction) containing the 75 protein molecules needed for regeneration the mathematical probability is 10 to the 9700th power.

5. Plants are able to make and store all standard amino acids; humans are not. Out of the 20 amino acids required to structure a protein molecule, humans are only able to synthesize and store 10 (some authors suggest eight). The rest must be obtained from diet. Failure to obtain even one will result in body protein degradation as the human body is unable to synthesize those amino acids.

Scientists are able to understand the function of proteins by observing the three-dimensional structure and resulting stability of amino acid strands in a protein molecule; however, the complex process and information delivery system that makes this happen is still a mystery.

6. There is no such thing as a self-replicating, inanimate molecule, though evolutionists claim such a possibility existed eons in the past as a precursor to the creation of a living cell. As always evolutional proponents have

no qualms about suggesting metaphysical deviations from natural order when it suits their theory.

A cell can only reproduce itself into another cell when all the parts and instructions are present in the first cell. It cannot be built piecemeal because all its components must have been created instantaneously in order for it to exist.

The same is true for the first human body, plant or animal. Though obvious similarities exist in DNA codes (which evolutionists point to as proof of a common ancestor), scientists recognize that even slight differences in the organelles and cellular DNA produce different life-forms. In all animal life, that original cell with its millions to trillions of components is instilled at conception with its distinct coding and the unique parts necessary to reproduce, supply the proper balance of oxygen, and grow into only that which its original DNA coding directs.

DNA coding is transported to long chains of amino acids, and most protein molecules contain one thousand or more amino acid chains. At the molecular level it is observed that these protein sequences are nontransitional. Simply stated, amphibians cannot genetically evolve into mammals. Basic types (fish, amphibians, reptiles and mammals) are isolated from one another. Crossover between kinds is precluded at the molecular level.

1. Evolution is impossible if crossover is impossible.
2. Evolution is impossible if a common ancestor is impossible.
3. The fossil record supports kinds by design not slow evolutionary changes (mutations) in kinds over time.

Evolution is a myth. It never explains the origin of organic life from non-life, much less the variations and complexities of life and certainly not the origin of reason, except to say it evolved like the rest of their aberrations ("deviation from truth"). Most people do not bother to look beyond the fabled "monkey to man" myth to question the more complicated and tedious aspects challenging the theory of evolution. The complexity of only one living cell demands a creator God. Man has never duplicated that miracle, why should one believe that random chance might do so?

Henry Morris, author of "What is Creation Science?", observed that in order to hold to the evolutionary model in the face of scientific rebuttal and common sense would indicate something else other than science is the persuasive argument.

"How great the love the Father has lavished on us, that we should be called children of God! And that is what we are! The reason the world does not know us is that it did not know Him" (1 John 3:1)

The myth of evolution has so infiltrated our schools, our culture and the way the world views God and His creation that Christians need to reconsider what message they are giving the world by its acceptance or their silence. Perhaps we have lost control of our schools, but surely we have not lost control of our churches.

Faith is only a virtue when it is grounded in God's truth. While sinful man is delighted to accept and promote a godless myth, Christians should remember that in order to be set free by the truth they must first know the truth and then believe the truth.

"If you hold to my teaching, you are really my disciples. Then you will know the truth, and the truth will set you free." (John 8:31-32)

Vestigial Structures, Pseudo Genes and Retroposed Gene Copies:

Allow me to begin this short subject with, I trust, a simple definition of each header.

1. Vestigial structures is an old evolutionary argument proposing that some visible evidence of physical structures that no longer are functional in an animal are proof of a former common ancestor and, therefore, proof of evolution (i.e. functionless eyes of blind cave-dwelling animals, the apparent pelvic bones of some snakes and the suggested gills and a vestigial tail on human embryos).
2. Retroposed gene copies have been observed alongside or behind ("retro") the original DNA gene. They appear to be ("posed") an exact copy of the original gene, functional and non-detrimental.
3. Pseudo genes or false genes are genes that appear to be not fully formed according to the parent gene and float randomly within the DNA molecule. One scientist projected that there might be as many as 1.1 percent of pseudo genes in the human body.

The implication is clear. What vestigial structures failed to prove, molecular DNA pseudo genes now sets about to prove or, at least, imply.

While some functionless pseudo genes and retroposons appear to be discarded over several generations, others are passed down. It also appears that the same pseudo genes and retroposons often appear in different animal classes, indicating to some evolutionists a common ancestor.

Here's the problem! Shared DNA among all animal kinds is understood by scientists. Why would shared pseudo genes and retroposons not be expected? Molecular design must be similar in order for life to survive in a common environment. The same is true of all the forces of nature, including gravity and the common exchange of oxygen and carbon dioxide between plants and animals. All must exist in concert or all perish.

Pseudo genes are not evidence of a common ancestor, but rather, evidence of design. DNA similarity is a must. If pseudo genes are the mechanism for new species, where are the new species? There should be thousands if not millions of different and yet similar kinds of animal life alive today or in the fossil record. We are not talking about different breeds of bears or dogs, which are only slight variations of kind and permitted through natural (within kind) reproduction. We are talking about the many required crossover (transitional) stages between amphibians and mammals or reptiles and fowl. Birds did not evolve from reptiles as most evolutionists suggest. Reptiles are cold-blooded creatures and birds are warm-blooded with hollow-like bones.

Over millions of years humans should have evolved into thousands or more human body types, and some still with tails and body hair. Every set of parents would potentially be the breeders of a new human-like race. If a common ancestor between humans and apes is to be considered plausible, and pseudo genes are responsible, then thousands of half breeds, quarter breeds, etc. should exist.

According to God's Word, this world in its fallen state will not last forever. We have already discussed several natural realities to confirm that statement, including the decreasing magnetic field above the earth and the second law of thermodynamic. Pseudo genes and any possible resulting mutations are quite possibly one more example of man's rebellion against God, his earthly death sentence and decay. Mutations degrade, they do not enhance. Man no longer lives in the "garden" perfection. Mutations do occur within limits; however, God's plan of redemption runs according to His perfect design and omnipotent control.

Mutated changes caused by a suspected variation of DNA sequencing or pseudo genes have never been shown to be advancements in kind or species. They have always proven detrimental in the short run, as countless cases of inbreeding have shown.

What we have discussed, however, is the presence of pseudo genes and retroposed genes that have never been shown to change the expected DNA program of any species. In other words, their presence or absence makes no difference to the DNA recipient embryo, and in many cases they disappear after several generations. They neither provide proof for a common ancestor nor evidence of natural selection.

In the next chapter we will consider some of the apparent corroborating evidence used by evolutionists to demonstrate an ancient earth. I suspect that faith in chemical dating has been the most persuasive argument many Christians have accepted while denying a biblical six-day creation. In fact, one might say, they have joined the congregation of radiometric dating.

Chapter 12

Carbon and Radiometric Dating

Lest you doubt the intent of secular evolutionists, read the introduction to an article in "National Geographic" entitled "Prehistory."

> "When compared to the history of humankind, let alone that of the earth, the inquiry into the development, roots and relations of humans is very young indeed. Up until the 18th century, the biblical story of human creation—'So God created man in His own image, in the image of God He created him; male and female He created them'.—was accepted as an incontestable truth in many parts of the world. Then, however, natural scientists—Charles Darwin, the most celebrated among them, appeared. They doubted the special status attributed to humans by the Bible and viewed their development with the context of a theory of evolution. The theory has since been supported and modified by the discovery of skeletal remains, primitive tools, and the remnants of ancient settlements. Their classification, dating, and evaluation using modern technologies has made possible an increasingly accurate perception of human origins."

Do you wonder where environmentalists get their passion for all living creatures? They "doubt the special status given to humans by the Bible." I am aware that a devout evolutionist is not going to suddenly see the light unless God shines that light; it is the Christian man, however, that needs a wake-up call in order to see the dangers of the religion of evolution.

The above premise that appeared in "National Geographic" is certainly one of the more naïve statements on evolution ever written—and there are many. The fact that this author believes that "skeletal remains, primitive tools and the remnants of ancient settlements" have anything to do with evolutionary descent is interesting. Yes, we have discovered all of these; however, none of them came with birth certificates, patent pending or construction dates.

Evolutionary Assumptions

Thus far evolutionists have made several general assumptions:

1. The earth's strata were caused by either slow erosion or many short-term catastrophes with long periods of quiet.
2. Early fossil discoveries appear to be in descending, physically more complex order—worms and smaller life-forms at the bottom and mammals and larger life-forms at the top. Evolutionists refer to this as the "Principle of superpositions", whereby, they are able to affix a "relative time scale".
3. Life began with a common ancestor and regenerated again and again through a process of gene mutation and natural selection to ultimately produce all plant and animal life-forms.
4. The earth's plate was once connected, allowing for animal migration; however, over long periods of time it became separated by seismic activity into seven continents.

5. One can date the age of fossils based on their placement in the Geologic Column, and one can date each geologic stratum based on its fossil content.
6. Long periods of time are necessary in order for life to evolve; therefore, the earth must be very old.
7. Animal and plant life began by spontaneous generation (life from non-life) by means of an unexplained energy source and continued to evolve through mutations, environmental influences and random circumstances not presently known.
8. According to secular evolutionists, biblical teaching of creation by design over a six-day span is religious myth.
9. Though gaps exist in the evolutionary theory, including no intermediate life-forms, they can be explained by anomalies, or will one day be explained by future discoveries.

Pulpit Reluctance

Why are many pastors unwilling to preach the Genesis creation inerrancy? Why are they unwilling to say evolution is 100 percent wrong and that God created the universe and all life according to the only way it could have possibly begun — that the Genesis account is correct and Darwin's assumptions are incorrect?

My assumption is that many pastors have either accepted or capitulated to evolution as fact and have adjusted their biblical interpretation and their preaching accordingly.

Do you wonder why some pastors do not preach with conviction? Perhaps they are not convinced of biblical inerrancy. If the biblical record of creation is a myth, what else is subject to human alteration, re-evaluation and humanistic re-interpretation, all in the name of theological progressivism?

Why is it so easy for our children to be spoon-fed a steady diet of evolutional assumptions when no real scientific proof has been forthcoming—only an unscientific theory? The answer may lie in the fact that the institutional church has been unwilling to stand on the entire Word of God, especially that Word that brought the world into existence.

An Aside

Few Southern Baptists are aware of what took place in Louisville, KY, at the Southern Baptist Theological Seminary in 1879 as it related to evolution. This is the same seminary from which I graduated in 1970.

A highly respected Hebrew professor, who shared a close relationship to Lottie Moon (there was talk of marriage), became interested in this new science of evolution. Born in Norfolk, VA, (my birthplace also) in 1866, Crawford Howell Toy traveled to Berlin in order to study this theory more closely. He returned a believer in evolution and, as a result, his entire view of inspired Scripture was altered. He later joined the faculty of Harvard University after being dismissed from Southern Seminary.

The problem, as I find myself stating over and over again, is that once the inerrancy of Genesis is challenged, all other scripture and biblical miracles are open for re-interpretation.

It is said that Lottie Moon broke off the relationship based on religious differences and, soon thereafter, began her missionary work in China.

The Dating Game Continues

Carbon dating differs from radiometric rock dating in several significant ways. Though many early attempts were made to use carbon dating to project millions of years into the past, no serious attempt is used today for fossil samples suspected to be older than 50,000 years; however, with radioisotope and radio isochronal dating of rock samples, some scientists still attempt to extrapolate to millions or even billions of past years based on atomic chemical reactions that take place in rocks.

Carbon dating is used only on formerly living fossilized or non-fossilized plants and animals (wood, plants, tree limbs, bones, etc.). Radiometric dating is most often restricted to igneous rock formed when hot molten material cools and solidifies (granite or basalt).

Carbon Dating

Carbon 14 is created in the atmosphere when the cosmic radiation of the sun blocks neutrons out of the atomic nuclei of Carbon 12, causing them to collide with nitrogen and convert into carbon 14. Carbon dating is based on the observed displacement of Carbon 14 atoms initiated within a plant or animal once it dies.

Once Carbon 14 has been produced, it combines with oxygen along with Carbon 12 into carbon dioxide and is absorbed by plants. Animals eat the plants or plant byproducts and then people eat the animals and the plants, ingesting a measurable amount of C12 and C14 in their bodies. It is expected that the ratio between C12 and C14 will remain constant until death. Today's ratio is one C14 atom for every one trillion C12 atoms.

As soon as the plant or animal dies, Carbon 14 begins to decay, changing back into nitrogen. Scientists say it will convert half of its total atomic weight every 5,730 years. An additional one half of its remaining Carbon 14 weight will convert in the next 5,730 years. Carbon 12, as a stable element, will remain constant with the existing atmosphere. As Carbon 14 decays, an increasing ratio between Carbon 12 and Carbon 14 develops—the greater the ratio, the older the sample.

Based on present C12/C14 ratios in the atmosphere, a fossil or a wooden table leg can be theoretically dated and an age affixed according to its present C12/C14 decay ratio.

In theory, carbon dating sounds reasonable; however, projections into the past require assumptions that are not observable, therefore, not scientifically verifiable.

While decay and the ratio created by decay tend to be accurately measured, projections into the past based on today's atmosphere as well as other unknown variables create too many untrustworthy conclusions. This, according to one physicist, is another example of evolutionists creating "a body of opinion, speculations and methods for interpretation of observational facts so that they fit into the philosophy of naturalism."

Variables

1. Some plants discriminate against carbon dioxide containing Carbon 14. Some take up more and some less. If the initial ratio is unreliable, then the age projection will be as well.

2. The ratio of C12/C14 in the atmosphere has not been constant. Massive burning of fossil fuels during the industrial revolution resulted in less C14 in the atmosphere.

3. Atomic testing in the '50s created more C14 in the atmosphere.

4. Scientists over the past 35 years have made scientifically monitored findings related to the magnetic field that surrounds the earth. Their findings suggest that based on the observed decay of the earth's magnetic field of 7 percent over the past 130 years or as much as 10 percent over the past 150 years, the indication is that the upper limits to the extended age of the earth's magnetic field is about 10,000 years, if things continue on their present course. Evolutionists suggest this trend will reverse itself, though no explanation exists as to how this may occur.

> The amount of cosmic rays reaching the earth influence the production of C14 in the atmosphere, and the strength of the earth's magnetic field affects the amount of cosmic rays reaching the earth. A stronger magnetic field deflects the sun's rays; a weaker magnetic field allows for an increase in the sun's rays, thereby producing more C14.
>
> Now we know that the earth's magnetic field is not stable but is actually weakening with large fluxes of gamma rays appearing to be a common occurrence. Based on increasing C14 in the atmosphere, plants and animals that lived on the earth thousands of years ago, when tested according to today's atmosphere, will

have greater C12/C14 ratios, indicating a much older age.

5. Because the earth's magnetic field is getting weaker, thereby causing greater levels of C14 in the atmosphere today as opposed to past years, two important observations are in order as they relate to carbon dating:

> First, anything tested today based on today's atmosphere, higher in C/14, will always produce greater C12/C14 ratios and appear older.
>
> Secondly, if it is established that C14 degrades at a reliable and constant rate as well as the fact that, based on the earth's lessening magnetic field, as levels of C14 continue to increase in the atmosphere, what has now been projected by some is a formula for determining the limited age of the earth.
>
> Two variables are occurring—C14 is degrading at a measured rate and the radiation of the sun is causing greater levels of C14 in the atmosphere. At some point these two variables will equalize—a point of equilibrium. If the earth were created today it would take approximately 30,000 years to reach that point of equilibrium where C14 is being degraded at the same level it is being created in the atmosphere, thus limiting the age of the earth to less than 30,000 years, as that point has not yet been reached.

6. A Genesis flood would greatly affect the ratio between C12 and C14. Until the earth was able to replenish plants that had been buried in the flood, anything that died during the

flood could not be reliably tested based on C12/C14 ratios today. Six-day creationists suggest that, based on their recalibration, fossils carbon dated 35,000-45,000 years old could be easily recalibrated to the year of the flood (approximately 2400 B.C.), if the flood's massive chemical effects on the atmosphere are considered.

7. One medical scientist, having performed significant research on Carbon 14 dating, states that fossils originally dated according to the Geologic Column at 350 million years ago, when carbon-dated with a zero age assumption, should be dated at 55,000 years old. This recalibration makes them compatible with a date as low as 4,000 years ago when the effects of a worldwide flood are considered.

During the 1950s Dr. Willard Libby developed the technique of radio carbon dating as a way of verifying historical or pre-historical dates (before recorded events). Though he was committed to the evolutionary theory, he never suggested that his formula could be used to extrapolate dates beyond a few thousand years and/or recorded events.

> The C14 isotope is tested in a sample with a mass weight of less than 1 gram (.352 of one oz.). The sample, of course, might be an entire leg bone; however, in order to determine the rate of Carbon 14 decay, one must isolate that element and compare its decay ratio with C12 based on today's atmosphere, which presently contains .0000765 percent of C14.
>
> Radioactive C14 can be detected with a Geiger counter. In today's atmosphere each gram of C14 should produce approximately 16 clicks per minute.

A sample that is calibrated to be 5,730 years old (the presumed half-life) should produce eight clicks per minute or four clicks per minute if the sample is calibrated to be 11,460 years old.

The scientific counting of radioactive decay in C14 atoms is not the issue. Several methods are employed and aided by an Accelerator Mass Spectrometer (the use of refracted rays of light whereby wave lengths are measured). Once the amount of C14 decay is established, it is compared to C12 and a ratio between the two is established. It is the creation of a "calibration curve" at which point trouble begins.

The following represents the expected rate of decay based on number of years:

Numbers of years in age	Observed Decay
5,730	.50 percent
2,865	.25
1,432 ½	.12 ½
716 ¼	.06 ¼
358	.03125
179	.015
89	.0085
44	.004
22	.002
11	.001
5 ½	.0005
2 ¾	.00025
1.4	.00012
8 months	.00006

Untrustworthy Variables

   a. Even when historically verifiable dates are used in order to calibrate observed C12/C14 decay ratios, that does not account for changes in the atmosphere upon which those calibrations are based.
   b. More dubious than that is when calibrations are made based on geologic locations assigned to be a certain age without a historical reference.
   c. Who determined that the rate of decay will remain constant?
   d. Who determined that our sample isotope has the same C14 content as all like samples unless all are tested?
   e. Does the unearthing and removal from its location affect the samples integrity?
   f. How is it possible for different isotopes from the same sample (either, fossilized or not) to yield conflicting C12/C14 decay ratios, which they often do?
   g. Ultimately, if the production rate of C14 in the atmosphere does not remain constant with the removal rate (primarily through decay), the ratio will change. Without a steady state of removal and production, starting amounts and accurate ratios are impossible to determine.

Because the beginning ratio is so important, most scientists today believe that carbon dating outside the range of recorded history cannot be used alone to determine age. Coal, supposedly formed millions of years ago from buried plant life, should have lost all C14 content; however, no coal that has ever been tested was without C14 levels, indicating a young earth and giving evidence to a catastrophic flood.

Because carbon dating is directly dependent upon the earth's magnetic field to maintain a controlled atmosphere and a reliable calibration curve, and because the magnetic field has changed significantly over the past 150 years, carbon dating has consistently yielded inconsistent dates. For example, the leg bone of a mammoth was dated at 15,380 years old while its skin was dated at 21,380 years old. Living mollusk shells were carbon dated 2,300 years old. Freshly killed seals were carbon dated at 1,300 years old. The shells from living snails were dated 27,000 years old.

Eleven of the earliest-known human skeletons were carbon dated 5,000 years old, while bones from a supposed Homo erectus, which was originally geologically dated 250,000 years old, was later carbon-dated 27,000 to 53,000 years old. Dinosaur bones have been carbon-dated 20,000 to 34,000 years old instead of 60-plus million years old. For these and other reasons most scientists are reluctant to make exact claims as to the age of a sample based solely on the atomic decay of C14. In most, if not all, age projections they now seek an independent limiting factor, a verifiable second source.

Who Are You To Believe?

Allow me to make one observation at this point before attempting to explain radiometric dating of rocks. Who are you to believe?

As it always will be, faith, belief and/or agenda will ultimately determine who or what you believe. The point, however, is very clear. In America, especially over the past 60 years, we have all been indoctrinated to believe that all these evolutionary claims are fact—that they have been proven. The reality is 100-percent opposite—they have not been proven and what evidence that has been suggested is highly suspect and completely unverifiable and, therefore, not proof.

In the Scriptures we are told how on many occasions Christian men of old spoke with boldness the gospel of Jesus Christ. If for no other reason, when the straw dog of evolution is logically and scientifically challenged, Christians should be encouraged to speak more boldly of their faith.

"Science is limited to the present . . . no matter how sophisticated the scientific equipment, it will never allow a person to analyze data other than in present time." (Holman Illustrated Bible Dictionary)

Radiometric Dating of Rocks

While most people who have at least heard of carbon dating assume that it is the source or proof for a multimillion-year-old earth theory; in fact, that is not the case. One of many science textbooks I read boldly states that "modern dating techniques have provided the range of dates attributed to the geologic time scale"—as if all are uncontested. In fact, it is the geologic time scale (Lyell's Geologic Column) that has influenced carbon and radiometric dating.

For evolutionists the hope was that radiometric dating and its decay process in certain radioactive elements, also based on half lives, might be observed and the age of rock samples from different strata containing those elements might be verified. However, radiometric dating is subject to many of the same restrictive variables as carbon dating.

Chemists explain that the radiometric decay process might best be understood as a mother-daughter relationship. For example, potassium-40 decays to argon-40 with a projected life value of 1.25 billion years and rubidium-87 decays to strontium-87 with a half-life value of 48.8 billion years (the parent element decays to the daughter element). Evolutionists assume that the rate of decay can be extrapolated exponentially (without deviation) over long periods of time to millions and billions of years in the past. Several more mother-daughter elements exist, including uranium-238, which decays to lead-206 with a projected half-life of 4.5 billion years, and uranium 235, which decays to lead-207 with a projected life value of 704 million years. One might assume that with so many ways to determine the age of a rock sample, no doubt would exist as to its age, thus providing scientific proof for an ancient earth. Certain isotopes, however, from the same rock stratum have been dated millions of years apart. A lava sample from the recent Mount St. Helens eruption was dated as far back as three million years ago. The uses of different formulas for dating chemical decay have yielded vastly different ages.

The age of radiometric-dated moon rocks ranged from 10,000 years to billions of years with the same rock, though evolutionists quickly disclaimed the younger age. Almost as soon as this break-through was announced, discrepancies began to arise. Scientists are forced to make several basic assumptions in order to affix an age to a rock sample.

1. One must assume that no daughter element was present in the sample or estimate how much daughter element was present, and thereby project how much may have been lost before testing began.
2. One must assume that decay rates have been constant over the age of the sample and/or the earth. The

present exponential decay law is based on isotopes with half-lives of only 100 years.
3. One must assume that the sample has existed in a closed environment in which no additional element has been introduced into the sample by fractionation, slippage or mixing lines in either the mother or daughter element.
4. One must assume that the location of a sample in the properly named geologic stratum is an accurate starting age to begin testing and/or to establish a mother-daughter ratio calibration.

Scientists admit that they are not measuring the age of rocks; but rather, they measure the displacement of radioactive elements in isotopes. In order to affix an age to a rock based on the atomic change in an element, one must first determine a starting point; hence, the problem. The ones who measure chemical displacement are dependent upon others to supply the approximate starting age (back to the Geologic Column and circular reasoning), as a reliable ratio calibration.

If a rock sample is found lower in the earth's strata around fossils identified with a specific era that becomes the starting age. The gradient (degree or level) of displacement of the mother element into the daughter element is theoretically measured from that date and an approximate geologic age of the sample as well as future displacement ratios are established.

Scientists sought to solve the problem of single sample measurements by using the radio isochronal dating method, with which four or more samples were taken from the same rock unit and measured with the use of ratios and graphs rather than counting atoms in a single sample. Of course, they ran into the same problem of starting ratios and conflicting ages.

By using the isochronal dating formula, a group under the direction of Dr. Steve Austin removed a basalt (lava) sample from the oldest stratum in the Eastern Grand Canyon and another sample from the youngest stratum in the Western Grand Canyon. The geologically dated younger sample measured 270 million years older than the older sample.

The truth is that only if one factually knows the starting age of a rock sample is it even theoretically possible to determine the age of that rock through radio isochronal or isotopic dating. Rocks do not come with conception dates. They are formed by heat, cooling and pressure and can be formed over hundreds of years, not theoretical millions of years.

Another well-known example of misdating an isotope occurred following the eruptions of Mt. Nguaruhoc (Mt. Doom) in New Zealand in 1949, 1954 and 1975. Using the argon method on molten rock, dates ranging from 270 thousand to 3.5 million years were delivered. Excess argon rates were blamed on slippage from the upper mantle which, of course, is quite possible; however, if such occurrences are possible in rocks of recent origin why would one not assume that it might also occur in rocks dated millions or billions of years old? The reality is that rocks taken from lower geologic levels have often contained argon rates in ratio to potassium more consistent with higher geologic strata and many tests reveal ages that are too old, even for the evolutionary time scale.

Lava rocks from an 1801 Hawaiian volcano were radiometric-dated at 1.6 million years old. Rocks from a 1959 volcano in Hawaii were dated 8 ½ million years old. Mt. Etna in Sicily erupted three times between 1964 and 1980 and lava rocks from those eruptions were dated from 350,000 years to 2.8 million years old—all radiometric dated with a "zero" conception date.

In Australia, wood was found in lava rock discovered in the Tertiary stratum (geologically dated up to 62 million years ago). Carbon 14 residue was found in the wood sample, which should have completely disappeared in less than 50,000 to 100,000 years. The wood sample was carbon dated at 45,000 years old. Radiometric argon dating performed on the lava rock yielded an age of 45 million years old.

With the use of a mass spectrometer chemists are able to measure the decay ratio in a sample as small as a billionth of a gram with a margin rate of 5 percent or less. Scientific measurements and equipment are not the debatable issue. Even if every measured decay ratio tested 100-percent accurate, that would not establish a scientifically reliable age of a rock.

More Untrustworthy Variables

1. When lava rocks are hot, argon (which comes from the earth) does not escape; therefore, with potassium to argon decay measurements some argon would already be present before the atomic decay process began.
2. The flow of water is soluble to all mother elements as well as many daughter elements and would drastically change the chemical makeup of those

elements in a rock sample. Uranium 238 can be completely removed by water.
3. When magma from within the earth begins to cool following a volcanic eruption, both quick cooling (as in a flood) and/or no contact with air have a tendency to restart the clock and the daughter element is eliminated.
4. Some argon to argon measurements have provided negative ages where an individual might be holding a rock that has not yet formed according to radiometric dating.
5. Exposure to certain atomic particles emitted during radioactive decay (i.e. neutrinos and neutrons) as well as cosmic radiation changes elements. A large density of neutrinos would speed up decay.
6. Argon, radon and helium are extremely mobile and would likely escape if a rock were cracked.

Over and over various dating methods yield wildly different ages. The interesting aspect in all of this is that scientists, when using the "zero" daughter approach (where no assumptions are made as to a starting age or to how much of the mother element has already changed to the daughter element), quite often came up with dates consistent with the biblical record. Though, admittedly, we have also established that even when known samples of recent origin are radiometric dated, they are regularly misdated by thousands to millions of years.

The above-mentioned discrepancies in radiometric dating cannot be casually passed off as anomalies. Starting points are essential, though not a guarantee of accurate results and, if those starting points are determined by location in the

Geologic Column, we are back to the same circular reasoning. A chemical scientist will generally ask, "Where did you find the sample?" and adjust their initial ratio accordingly.

If one, from the outset, eliminates the biblical record of creation and a worldwide catastrophic flood, then alternate assumptions must be made. Whereas most evolutionists now acknowledge that perhaps many independent catastrophes occurred over long periods of time, including possible worldwide monsoons, they are determined not to consider a one-time biblical flood 4,000 years ago. Only the Bible is mythical—not their assumptions.

Yet to Be Explained by Evolution

1. The lack of erosion in the sediment layers of the Geologic Column, even those layers low in the column and subject to longer weathering conditions.

2. The bending and lack of uniformity in layering, which should not have occurred if the layers were slowly formed over long periods of time.

3. The fact that bent layers are not fractured, which indicates they were soft and pliable at the time of rapid formation.

4. The dinosaur bones, evolutionary dated over 65 million years ago, which still contain hemoglobin in the red blood corpuscles, soft connective tissue and the presence of DNA. None should exist a few thousand years after death. It is biochemically impossible.

5. The present rate of decay in the earth's magnetic field, which indicates an age of supportable life on earth at less than 25,000 years.

6. Based on accepted rates of atomic decay, many isotopes with a geologic age of millions of years in which the mother element is still present long after it should have decayed into the daughter element, which indicates a much younger age than what was expected from its location in the Geologic Column.

7. One of the most interesting questions to arise concerning radiometric dating deals with radio halos. As radioactive particles decay, they cause what is observed as halos (damage to the surrounding area of a rock). Any existing element present at formation of a rock will cause halo rings during its chain of decay from one element into another.

    What geologists noticed was that certain elements in a sample that were thought to have appeared via decay of the mother element had no mother-element halo. It was concluded that the rock must have formed quickly, capturing elements from surrounding sediment rather than over long periods of time, or that, on the other hand, there have been radical changes in past decay rates.

8. Evolutionists are reluctant to admit that glaring conflicts exist between radiometric dating and other forms of dating. One example is helium retention produced by the nuclear decay of uranium in zircon crystals commonly found in granite. Zircons existing in

nonevasive environments have been radiometric dated to be more than a billion years old, well into the pre-Cambrian era (600 million B.C.-plus). They were also measured according to helium diffusion at only several thousand years old. Helium diffusion and radioactive decay can both be rated with a high degree of accuracy in a lab; so why the vastly different results? Most geophysicists now suggest that it is impossible to accurately determine age based on chemical conversion without a historical reference point.

Because long periods of time are necessary to support a theory of natural evolution of life, the introduction of carbon and radiometric dating was heralded as the final straw that would silence six-day creationists. Always quick to disavow the existence of a divine creator, evolutionists believed they had at the same time proven the reliability of the Geologic Column.

Carbon and other chemical decay methods have proven to be inaccurate, contradictory and unreliable as a dating tool without a verifiable second source (recorded history). Those who set about to put exact dates on fossils and rocks fell into the same trap as earlier old-age proponents. It is impossible to apply dating into millions or even thousands of past years based on atmospheric and environmental conditions today; without an accurate starting point as a historical reference point, there is no way to estimate age based on atomic chemical displacement. Scientists may determine rates and ratios of change but not age; consequently, most scientists no longer try.

School textbooks will not explain to students the inconsistencies. The books continue to say that radio carbon and radiometric dating have provided evidence of an old earth; they generally avoid any conflicting findings.

A question that bothered me early in my understanding of evolution and atomic dating was that even if all the details of evolution and an old-age earth are not 100-percent accurate, what difference does it make? Evolutionists are suggesting millions and billions of years, not thousands as in the biblical account.

The sheer magnitude of multiple millions versus the biblical thousands seems to be overwhelming evidence that the biblical six-day creation and young earth are suspect. How can evolutionary science be so far off the mark?

Remember . . . we are not dealing with empirical or even observable evolutional facts. We are dealing with observable chemical displacement based on preconceived evolutional assumptions. If one's assumption is chance creation, then one's formula for slowly evolved life must be calibrated based on long periods of time.

Every test and every theory must comply with that assumption. When one test does not fit that formula, then it is discarded and replaced with a test that seems to fit. When one's calibration curve is based on millions of years as proposed by the Geologic Column, then the ratio curve must be calibrated to millions of years. There is ample chemical evidence that also indicates a young earth (in the thousands of years), though it is not given the same credence by evolutionists.

There is also no written record available to substantiate claims of an ancient earth. The basic evolutionary assumption is that "anything is possible if given enough time." Hopefully, you are beginning to realize that assumption is just not true. **Here are the debaters:**

1. **The Biblical Literalist** is one who believes in a literal six-day creation and that all Scripture is inspired by God and, therefore, trustworthy.

2. **The Biblical Progressive** does not believe that all Scripture is literally inspired by God but rather believes that some is allegory or myth and must be re-evaluated in light of today's world. They believe in a god confirmed by their own understanding and that the biblical account of creation, though containing allegorical truth, is a mythical story.

3. **The Evolutionary Literalist** is willing to accept any creation explanation that does not include God.

4. **The Evolutionary Progressive** admits to the inconsistencies of evolution and an old-age earth, the unreliability of evidence and the lack of scientific proof; however, this person doubts the existence of God and chooses to wait for additional human understanding and/or discoveries.

A Prominent Study

I would be remiss if I did not mention an extensive carbon and radio isotopic study performed by a team of seven scientists (geologists, geophysicists, physicists and meteorologists) led by Dr. Don DeYoung beginning in 2003. While Dr. DeYoung goes into great detail concerning his findings in his book "Thousands . . . Not Billions", I shall restrict my use of those findings by explaining his summary conclusions.

1. Carbon 14 dating must be considered unreliable when projecting past ages beyond historically verifiable dates.

    The presence of C14 in fossilized dinosaur bones, which should have decayed long ago but is now being discovered worldwide, scientifically proves that the Geologic Column is inaccurate as a dating tool; C14 residue has provided that evidence.

    In addition, the latent existence of C14 in coal deposits evolutionary dated hundreds of millions of years old and also found in diamonds dated millions and sometimes billions of years old is strong evidence for a young earth as well as a catastrophic flood.

2. Zircons are tiny crystals that often occur in granite, one of the most abundant rock types on earth with a radio isotopic date of over 1.5 billion years. Helium atoms, however, which result from the decay of internal uranium atoms, should have long ago escaped from zircons . . . and that is not the case. High

concentrations of helium are still found present in zircons.

These scientists conclude that based on the lack of helium diffusion in zircons the age of the earth is closer to 6,000 years.

In addition, the tracking of helium atoms indicates a point of accelerated nuclear decay in the past, providing evidence of a catastrophic flood.

3. Radio halos, those tiny spherical circles left by the decaying mother element in a rock sample, are regularly found in uranium and polonium sample isotopes. When no longer shielded from atmospheric exposure, polonium decays quickly, within hours or days; therefore, when many radio halos are discovered in rocks dated as far back as 1.7 billion years, one expects to find the parent (uranium) element clustered alongside the daughter halo (polonium). Evidence for the parent, however, is often missing.

These scientists concluded that apparently the motherless element (polonium) was rapidly removed from the uranium parent during an underground catastrophe.

4. Finally, this team of scientists also points out that radio isotopic dating methods are forced to make several assumptions that are not verifiable:

    A. Even though the chemical condition of a rock sample can be accurately determined, the assumption of a conception date is challenged by

many discordant dating results including ancient dates ascribed to recent lava rocks.
B. The assumption that one can always determine the open or closed nature of a rock sample is unverifiable.
C. That nuclear half-lives have remained constant throughout the past is challenged by the presence of helium in zircons.

Extensive nuclear activity occurred twice during the Genesis account—first, during creation, and finally during Noah's flood. Though Dr. DeYoung suggests his work must continue, he and his team believe ample chemical evidence exists to indicate a recent creation and a catastrophic flood.

With all its inconsistencies, many are still determined to rely on radiometric dating as proof of an old-age earth. The truth is that there are as many chemical examples of a young earth, including the amount of helium in the atmosphere, a decreasing earth's magnetic field and others we have also mentioned. All chemical dating methods are subjective; to claim that they have provided proof of an old-age earth is simply not possible.

One might say, "That's enough proof for me", and that is one's choice. Remember, however, the geologic column and radiometric dating is driven by those who reject six-day creation, not by empirical science. If one were to add 2 + 2 and the answer was sometimes 4 or sometimes 4,000, they might question the method of calculation; however, if one was determined to prove an ancient earth, they might throw out the 4 and, based on future discoveries, extol the 4,000—it is their choice.

Since the time of Darwin a great deal of new evidence for and against evolution has arisen. Next we will discuss the basic evolutional doctrines, along with some objections to evolution as well as the work of a well-known evolutional theist, Dr. Francis Collins.

Chapter 13

The Evolutionary Point of View

> "As we have seen, one of the strongest evidences of evolutionary descent comes from the fossil record, which presents several examples of evolutionary transitions from one class of organisms to another. In addition, the fossil record grades clearly and unmistakably from simple early life-forms, which appear early in the geological column, to larger and more anatomically complex forms, which appear later. The sequence of the appearance of various fossil groups—first invertebrates, then simple vertebrates, then jawed fish, then amphibians, then reptiles, and finally birds and mammals—is exactly what we would expect from evolutionary descent with modification, with the organisms appearing higher in the geological column being the modified descendants of those organisms which appear lower in the column." (Copied)

Well-stated as it is, and exactly as it is taught in the classrooms of children all over the world, this belief is simply not true. I would request that you reread the above paragraph one more time, for it is the heart and soul of evolution. Okay? Now, let's examine what is being said.

Quote one:

"One of the strongest evidences for evolutionary descent comes from the fossil record . . ."

The truth is that the supposed "fossil record" is the only supposed "evidence" of evolutionary descent. Where or what is the other evidence? Many scientists say that a DNA shift in organisms significant enough to create a new kind of life-form is impossible. Either position is based on opinion and not verifiable proof; as such a shift has never been demonstrated by science to occur. And we are not talking about smaller and larger dogs. Evolutionists are suggesting that with enough DNA tampering (mutations) over long periods of time, jelly fish and seaweed became dinosaurs, tigers and pine trees or fish, frogs and cattails (complexity built upon complexity). The entire concept is ludicrous scientifically. But because evolutionists claim to envision a significant pattern of lower to, what they assume to be, higher life-forms in the apparent earth's layers (the Geologic Column), none of which is uniform as they imply, their only possible conclusion, and according to Darwin, is that all life descended from a single ancestor and evolved by means of random, natural selection and survival of the fittest life-forms.

Consider for just one moment. If that were the case and, say, 80 million years existed between one life-form and its evolved survivor into a new life-form, then two things must occur:

1. There would be many temporary, "trial and error" (transitional) life-forms readily apparent in the theoretical fossil record between the "fishes" that are gradually turning into turtles or beavers or birds. There should be many weird creatures discovered in the fossil record, now extinct of course, but demonstrating a clear line of transition from something to something else (a fish with legs perhaps). These weird and extinct transitional creatures should be far and away the most

discovered and greatest publicized of all fossils. Why the most publicized? Because evolutionists would then possess a clear linage of something gradually turning into something else, as the old something else passes out of existence (fails to survive); and as yet, none have been discovered.

The problem for evolutionists is that the massive amounts of fossils that have been discovered are of those species presently alive in their same form or now extinct in their same form. The exact opposite should be found in the fossil record. Evolving species should vastly outnumber those species in their present form if all species are evolving over long periods of time.

We are not talking about a few required DNA mutations necessary to turn a frog into an alligator, separated by only six chromosomes. We are talking about thousands if not millions of complex DNA mutations necessary, thereby, creating a massive transitional fossil record between any and all life-forms.

2. We must be constantly reminded that evolutionists are talking about multimillions of years between these very small to very large creatures. If, as evolutionists insist, each succeeding generation gradually, and I mean gradually, turned into something new, which is the only possible evolutionary model, then all lower-level life-forms should be extinct and the higher-level life-forms should represent the fittest survivors, alive today or now extinct. And will someone please explain to me how the chicken survived as the fittest under that formula? Neither should there be sheep in today's

world. Were it not for the evolution of farmers and shepherds, helpless chickens and sheep would be extinct.

In addition, any exception to that slow and gradual rule is impossible according to the evolutional theory itself. Heaven forbid that we should find dinosaur bones at lower levels or brachiopods at higher levels of the Geologic Column. Why? Because they are separated by hundreds of millions of years of evolution, it should be impossible for a creature of 60 million B.C. to be discovered alongside creatures in the 480 million B.C. stratum. How could such a thing happen as all life-forms struggle to survive into the next life-form?

Well, guess what? They are found all the time. Creatures supposedly separated by millions of years buried together. There should be no exceptions to the evolutionary theory of Uniformitarianism (slow erosion of the earth's surface over millions of years, burying existing generations of evolving life-forms beneath layers of rock).

Quote two:

". . . the fossil record, which presents several examples of evolutionary transition from one class of organism to another."

This is simply not true and the one who spoke it should be spanked. I'll address only two quick points here.

1. A simple sutured ammonite (a coiled, chambered mollusk, now extinct) is not a transitional life-form to a slightly more complex ammonite. They are the same

kind just as a tiger is of the same kind as a lion. Both ammonites lived together and became extinct together. Variation of kind is not transitional.

Others have pointed to the platypus (a semi-aquatic monotreme who, like a mammal, gives live birth rather than having eggs) as a transitional life-form. Though possessing traits of a mammal, a platypus has a venomous barb on its hind leg (the male platypus), a duck-like bill, a beaver tail and otter-like feet. So what is it transforming into—a beaver, a duck, an otter or a stingray? And by the way, the platypus still exists alongside the beaver, the otter, the duck and the stingray.

The Archaeopteryx is considered by most evolutionists to be the first bird (evolving from a reptile dinosaur to a feathered bird). Approximately eight fossilized track specimens have been discovered of this creature and all have been geologically dated at 150 million years ago.

It measured one-and-a-half to two feet in length and, except for the long, stiff, bony tail as well as a dinosaur-type jaw with sharp teeth, it very much resembles a bird. Few believe that it could fly, although some suggest it might glide.

Later examination has led some to believe that the "Berlin" (discovered in 1887) and the "London" (discovered in 1864) specimens (the best known specimens) are fakes. Part of this assertion comes from the fact that in some of the photographs and engravings the feathers have a curved appearance, whereas, in others they appear straight. Other differences between the specimens are also noted, which has led some to speculate whether, in their enthusiasm to confirm Darwin's theory, artistic imagination was employed.

There is no way for evolution to prove its claim of transitional creatures between reptiles and birds. However, certain realities do exist: (1) Genetic scientists will tell you that DNA crossover today is not permitted between reptiles and birds, though some continue to insist it might have happened millions of years ago. (2) That which appears in textbooks as an evolutionary tree based on visual similarities between kinds is not based on biological relationship. Legs do not evolve into wings or scales into feathers. (3) There were many flying, non-bird reptiles with teeth and claws during the time suggested for the dinosaurs by evolutionists. The ostrich has reptilian-type claws on its feet. Though it probably existed, no compelling reason exists to assume the Archaeopteryx was a crossover or transitional life-form.

2. For most scientists the "Cambrian Explosion" (approximately 580-560 million evolutionary years ago), during which all present or now-extinct life-forms suddenly appeared, as observed in what evolutionists refer to as the Cambrian stratum, completely destroys the slow and gradual transition of plants and animals. Of course, evolution has no answer as to how this could have happened.

Quote three:

"The fossil record grades clearly and unmistakably from simple early life-forms, which appear early in the Geologic Column, to larger and more anatomically complex forms, which appear later."

Darwin could not have said it better back in 1859. That of course, is the hope and not the reality. The fossil record is relatively shallow and jumbled. Smaller life-forms are more commonly found at lower regions of the earth and larger animals and plants at higher levels even today. Why would that not be generally expected in nature?

Ken Ham in his recent book, "The New Answers Book #1", published in 2006, reminds us that 95 percent of all discovered fossils and fossil imprints are "shallow marine organism, such as corals and shellfish," and "95 percent of the remaining 5 percent are algae and plants." What that means is a small record of vertebrates and invertebrates with the vast amount of those being fish and insects. Mammals, including dinosaurs and humans, make up a very small portion of the fossil record and most of those are post-flood and ice age. In fact, the fossil record represents exactly what might be expected from a catastrophic flood.

Rather than repeat some of the material discussed in other chapters, we do need to be reminded that large does not necessarily mean more complex. It simply means large. A single cell may have a trillion or more microscopic components. All life-forms are large in cellular complexity. "Lower to higher life-forms" is an irrelevant misnomer. For any life to exist it must share a similarly complex molecular structure as all other life—be it plant or animal, small or large.

Chromosomes, one of the most complex organic particles of matter in the universe, might indicate the expected evolutionary progression of a less complex life-form to life-forms of greater complexity. Under that scenario, fruit flies with eight chromosomes would be the ancient ancestor of humans with 46 chromosomes. And assuming man continues to evolve, he may one day become a chicken (78 chromosomes) or a goldfish (94 chromosomes) or a fern with 480 chromosomes.

What evolutionists would like one to believe is that genetic complexity also evolved, thereby allowing smaller beings to become larger and more complex beings. That is impossible. An ant must have the same degree of cellular complexity as an elephant. Neither could exist were it not so, as neither could exist were they not created independently.

The clam is a typical bivalve mollusk. Though there are around 11,000 variations of the so-called clam, including fresh water mussels, scallops, marsh clams and giant clams, all are generally identified as bivalves. The clam has no head and generally no eyes and yet it has a functioning kidney, heart, mouth and anus. Its open circulatory system transports oxygen and nutrients to its organs, and it eats through a filtered feeding system.

Research has recently confirmed the presence of RNA centrosomes (protoplasmic substance responsible for cell division) in the clam, which guides the translation of genes into proteins. The eggs of certain surf clams have been used as research models for similar versions of human cells.

Though DNA varies in all life-forms, cellular complexity as well as unique structural complexity is an absolute in all animal and plant life. The messaging found within the cell nucleus of a tiny amoeba is equal to that of a 30-volume Encyclopedia Britannica; the DNA messaging of the entire amoeba cell would fill 1,000 encyclopedias.

The Holy Geologic Column

There are few defensible positions based on observation that evolutionists can claim as evidence to support their belief in an ancient earth and gradual descent of life. Radio carbon and other chemical dating is untrustworthy beyond recorded history as there is no legitimate way to determine a starting age from which to measure chemical decay. When a "zero" starting age is used, dates consistent with the biblical record are often reached. In addition there is no way to confirm the integrity of a sample isotope or reliable evidence that demonstrates today's atmosphere is comparable to the atmosphere of thousands, much less millions, of years ago. Finally, there is no way to extrapolate exponentially to 10,000 or 10 million years based on short-term isotopic decay ratios (where one element gradually surrenders to another element).

In addition, fossilization of organic plants or animals is not an indication of extended age as mineralization may occur over a relatively brief number of years.

What's left is a theory supported by discordant rock layers and jumbled fossils between rock layers. That's it—nothing more—the holy Geologic Column. In the face of hundreds of unnatural, illogical and mind-boggling exceptions, evolutionists remain determined to pay homage to a shallow fossil record in the holy Geologic Column.

Noah's Adventure

It is to be expected that evolutionists would challenge Noah's flood, as this catastrophic flood, if verified, would destroy all that's left of evolution by destroying what's left of the Geologic Column.

When God closed the door behind Noah's family and "on that day all the springs of the great deep burst forth, and the flood gates of the heavens were opened" (Gen. 7:11), this was not a regional river that overflowed its banks. The flood was accompanied by volcanic eruptions to which the aftermath of lava flows all around the world testify. Earthquakes and undersea volcanoes prompted great tsunamis. Some suggest that a comet may have struck the earth simultaneously, altering its axis and creating a brief ice age to which there is also compelling evidence. Others suggest that an ice age was precipitated by the flood and created bridges to locations on the earth now separated by water. God's purpose was judgment and destruction, and it was thoroughly accomplished.

For those who are not blinded by eyes determined to remain godless, the evidence of massive destruction is all around the world. Huge gorges, jagged mountain ranges and isolated rock formations are testimonies to massive flooding. Compare the topography of today's earth to slow and gradual erosion over millions or billions of years. Is that the picture you see in the Grand Canyon? Are the layers smooth and uniform as they must be based on millions of years of slow erosion? The jagged surfaces of the earth shout the mighty power of God; the complexity of a single cell reveals His omnipotence.

Over and over again I have read the contemptuous way angry evolutionists speak of those who accept God the Creator and a

Genesis flood. "Extreme conservatives" and "Bible college fundamentalists" are some of the kinder references, as if no one who does not believe as they believe has ever attended a class in high school biology or lives on the same intellectual plane as they do.

The reality is that the more the light of evidence and logic is shed on the theory of evolution (which has taken 150 years it seems), the more highly educated scientists are now beginning to recognize and finally admit its many weaknesses. Many of those who, unfortunately, refused to acknowledge a supernatural God are among those who are now rejecting evolution as a valid theory of creation.

It is difficult for a person of science to accept the premise that life mysteriously created itself from non-life or how order might arise from chaos. They wonder how it is possible that by violating all the known laws of nature a cell more complex than anything man could ever imagine much less duplicate might randomly appear. They wonder what unknown gene gave man, and only man, conscious awareness and the power to reason. It is with that same reasoning they also question how all kinds of life might evolve from that single magic cell (a common ancestor)—no matter how close DNA sequencing might be among species.

What's Left Standing?

Hopefully you have not arrived at this point without realizing what's left standing. From the beginning of this study we acknowledged that creation by evolution or by design was not scientifically provable because neither theory is able to be scientifically replicated.

That certainly does not mean that creation must forever be a mystery. Once again, remind yourself that only two possible ways exist that the universe and life on earth could have begun:

1. Accidentally (based on extreme anomalies in Natural Law over long periods of time) or,
2. Designed (instantaneously spoken into existence by a power outside this natural universe).

Purposefully Fashioned by God

"The earth is the Lord's, and everything in it, the world and all who live in it, for He founded it upon the seas and established it upon the waters." (Ps. 24:1-2)

"He [Jesus the Christ] was chosen before the creation of the world, but was revealed in these last times for your sake. Through Him you believe in God who raised Him from the dead and glorified Him, and so your faith and hope are in God." (1 Peter 1:20-21)

Evolutional Theists

One of the most difficult areas of debate for a biblical conservative to enter is the conflict between those who believe God was involved in the evolutional process and those who believe God created each species separate according to the Genesis six-day account. Based on a fairly recent poll taken in 2004, as many God believers in America who believe God was involved in the evolutional process exist as those who believe that God created life according to the biblical account.

Francis S. Collins, author of "The Language of God" and nominated by President Obama to lead the National Institute of Health, makes a compelling case for God's involvement in the evolutionary process. His extensive work in mapping the human genome qualifies him as an able spokesman for the evolutional theistic position. According to Dr. Collins, "No serious biologist doubts the theory of evolution."

While many biologists may fit that description, I might suggest that Dr. Collins check with Dr. Jerry Bergman, science instructor of Northwest State College in Archbold, OH, who has a Ph.D. in human biology from Columbia Pacific University, or with Dr. Henry Zuill, biology professor at Union College in Lincoln, NE, who has a Ph.D. in biology from Loma Linda University, or with Dr. Arthur Jones who has a Ph.D. in biology from the University of Birmingham and presently serves as a science and education consultant and member of the Institute of Biology in London, England, before he makes such a broad statement. All biologists are not in agreement when it comes to the theory of evolution. For the names of more serious biologists who believe in six-day creation, they can be found in the book entitled "In Six Days", edited by John F. Ashton, Ph.D.

The Case for Evolutional Theism

In April of 2003, approximately 2,000 scientists from around the world completed the task of mapping the 3.1 billion letters of the human genome. Among other things, it has been determined that only 1.5 percent of those letters are needed to code protein and that humans share a 99.9-percent similarity with other members of the human species and a low level of diversity with members of other species.

According to their findings, the gene sequencing that codes protein in both chimps and humans is identical and the random DNA segments between the two are 98-percent the same.

Dr. Collins' conclusion is that the Darwinian "Tree of Life", whereby all species originate from a common ancestor, can be "solely based on DNA sequence similarities." He believes that evolutionary mutations over long periods of time are the only scientific formula to explain all past and present species. He states that those who hold creationists views "are well-meaning, God-fearing people, driven by deep concerns that naturalism is threatening to drive God out of human experience" and "ultimately dilute the power of the scriptures to teach reverence for God to humankind . . . and put the believer on a slippery slope towards a counterfeit faith."

It is in the possible mutation of gene sequencing, though never actually observed, in which Dr. Collins bases his evolutionary assumptions.

Cytochrome-c is a protein gene product that consists of a chain of 112 amino acids, 19 of which appear in the same sequence in all organisms. It is the mutational substitution of the remaining 93, envisioned by evolutionists, that determines the creation of different species.

Human sequencing of cytochrome-c is the same as that present in chimpanzees; only one difference exists between humans and Rhesus monkeys—hence, the theory of a common ancestor creature. Differences in the 93 remaining amino acid sequences between humans and cows, pigs and sheep are 10, a horse at 12, a hen or turkey at 13, a fly at 25 and yeast at 44. It is primarily through the cytochrome-c gene sequencing that today's "tree of life" is constructed.

The question is, does the sequencing construction of cytochrome-c alone provide enough evidence to prove or even suggest evolutionary descent?

Except for the chimpanzee in which the amino acid sequencing in the cytochrome-c gene is the same, all other species differ from humans in at least one or more amino acid positioning and are only the same in 19 out of 112 with all species.

Geneticist Dr. James Allan, Ph.D. from the University of Edinburgh, states that "apart from the single gene controlling the constitution of cytochrome-c, humans and chimpanzees differ in many thousands of other genes." His conservative estimate is 5,000.

According to Dr. Allan, what evolutionists are assuming is that after generations upon generations of evolution, humans and chimpanzees have maintained the one cytochrome-c gene sequence in tact while the thousands of other amino acid gene sequences evolved differently. Based on that assumption, one is now to believe chimps and humans came from the same ancestor. He conservatively estimates the number of intermediary deaths required to produce a modern human under that scenario to be 150,000,000,000.

With so many intermediary deaths required to eventually produce a human descendant, Dr. Allan questions why none have been discovered in the fossil record.

Without giving any examples, Dr. Collins states that intermediate fossil forms have been discovered for birds, turtles, elephants and whales and that the second Law of Thermodynamics does not rule out evolution. He also states that decay rates for dating rocks have not changed since the creation of the universe.

Actually, unless Dr. Collins was present thousands of years ago and performed his own radiometric testing, no way exists for him to know whether decay rates have changed since creation. Questions and contradictions do arise when one attempts to exponentially extrapolate into the past based on today's geologic and atmospheric conditions, as well as the starting point of atomic decay necessary to suggest the age of a rock. In other words, unless one makes an assumption as to when a mother element began to decay into the daughter element (when the rock was assumed to have formed) or how much daughter element was present when it was formed or discovered, there is no scientific way to determine an age based on radiometric dating.

If, as the Bible tells us, a worldwide flood occurred 4,500 years ago, thereby causing rocks and rock layers to form quickly, then all radiometric dating must be calculated based on a starting point of 2400 B.C.; even then there is no way to know how much daughter element was present when those rocks were formed. But, of course, then it would not be necessary, because the Bible tells us when those rocks were formed.

While some evolutionists may insist that the Laws of Thermodynamics do not rule out evolution, if they are honest, they must admit that the Laws of Thermodynamics make a far more compelling and scientifically reliable case for instantaneous, six-day creation than they do for the yet unproven theory that preexistent nitrogenous chemicals were assisted by random heat fluxes, thereby overcoming the forces of entropy for a time in order to allow for the creation of the first cell. Later, stability is reestablished in order to begin a process of gradual, fortuitous mutational circumstances over long periods of time, where life must now overcome the formally suspended Second Law of Thermodynamics in order to survive and descend.

Who's Right?

As we have begun to understand, objectivity is not an absolute principle for some theistic evolutionists. The same can be said for some theistic creationists, though the reason for the lack of objectivity is vastly different. Both are examining the same evidence, reviewing the same natural laws and reading the same Bible; so why is there disagreement?

One can safely say that evolutional theists are also well-meaning, God-fearing people who sincerely believe that the cosmological and biological evidence for evolution is so compelling that those who argue against it are either misinformed or incapable of understanding the technical implications of recent discoveries—more specifically, as with Dr. Collins, DNA evidence. So what are the reasons for disagreement? Why can't we all just get along?

In spite of evolutional disagreements, Christians should be able to get along with other Christians. However, theological principles do exist that will, I suggest, continue to divide us into separate congregations.

1. The most important of these disagreements is evangelical: The teaching of the Gospel and scriptural authority. *"Euangelion"*, the Greek word for evangelism, means gospel or "good news." Christian evangelism is

the divine calling of people to respond to God's gift of grace and surrender themselves to God in Jesus Christ. Such a definition and belief system puts strict restrictions on how one reads and interprets Scripture.

The goal of evangelism is eternal salvation and communion with God through surrender to Jesus the Christ, not simply to teach "reverence to God." Apart from surrender to God's grace in Jesus Christ, there is no reverence for God. One cannot accept the authority of God in Christ by rejecting the authority of God's Word and God's designed plan for creation and redemption.

2. The Bible teaches that "sin entered the world through one man and death through sin" (Rom. 5:12). Evolution teaches that death entered the world first and that sin (at least in the minds of evolutional theists) entered only after humans evolved to their present state of conscious responsibility. If, as the Bible says, death is the last enemy to be defeated (see 1 Cor. 15:6), it makes no sense to believe that God would create life through a series of mutations resulting in the death of what has been conservatively estimated to be 150 billion evolving, human-like creatures over millions of years, nor does the Bible teach it.

3. Jesus taught that the days of tribulation (anguish and persecution) "will be days of distress unequaled from the beginning, when God created the world, until now—and never to be equaled again" (Mark 13:19). Evolution teaches that creation began by unequaled struggle, death and then continued to endure the

struggle for survival, cataclysmic activity (including volcanoes, comets, earthquakes and ice ages) throughout millions of years until the most recent days and the eventual evolvement of man.

Finally, if as the Bible teaches, a worldwide flood occurred 4,500 years ago, then the evidence evolution is so dependent upon—radiometric dating, the geologic column, a consistent, unchanged atmosphere and geologic uniformity—is completely destroyed. Not only does evolution deny a Genesis creation, it must also reject Noah's flood and the division of people following the Tower of Babel. All of those theories so important to creation of the universe, genetic mutations and the "Tree of Life" would be destroyed during a catastrophic flood.

God's Guiding Hand

Simply stated, because Dr. Collins is so convinced that the DNA evidence clearly allows for the possibility of mutated changes over long periods of time, he is compelled to adjust all other variables to fit his biological theory. He rightly envisions that God is able to see the beginning of creation from the end (which He is), but wrongly concludes that God would adjust His plan to fit an evolutional time scale. For Dr. Collins, and all other evolutional theists, God watched the universe and life create itself over billions of earth years and only entered that creation process at appropriate moments in the evolutional (death) process.

Dr. Collins is an admitted expert on genetics. He is the Francis Crick of the 21st century, but, Dr. Collins has a bias. He is not willing to accept that God is able to control the content of His Word. Dr. Collins believes that men wrote the Bible with little or no help from God. He must believe that in order to reject those passages that he feels are inconsistent with his own conclusions.

For those who accept the authority of God's Word, it is not acceptable to believe that evolution was controlled by "God's guiding hand", if evolution directly contradicts the biblical text.

Highly trained men of science, respected in their field of study as well as committed to a belief in the God of the universe, have looked at the same natural evidence and reached completely different conclusions concerning evolution and creation. While I choose not to question the faith of either group, what authority is one to base their faith upon if not God's Word? Paul Tillich said that "doubt isn't the opposite of faith; it is an element of faith." I see doubt as a Spirit-directed passageway to faith.

I never doubted the sin in my life or my inability to overcome that sin on my own. Based on the authority of Scripture, God's gift of His Son removed any doubt I may have had about God's love for me and His willingness through grace to forgive that sin. Since Christ entered my life and I availed myself to the power of His Holy Spirit, allowing me, as I believe, greater understanding and trust in His Word, doubts concerning the authority of Scripture have also been removed. I must admit all this is personal with me and that which overcomes my doubts and gives me peace. Even those Scripture passages I may not fully understand now, I do not doubt that God would reveal them with clarity if He so desired in order to advance His Kingdom.

For Francis Collins, the complexity of DNA molecules and the process of evolved life based on what he refers to as the "Language of God" give him "awe." For me and many others it is the simplicity of God's Word, spoken in a way that all generations might clearly understand and trust that gives me awe.

God has purposefully created and designed this universe. He has not operated out of chaos. The universe reflects a hands-on Creator. Humanity reflects God's sacrificial love beyond man's ability to fully comprehend.

"For this is what the Lord says—He who created the heavens, He is God; He who fashioned and made the earth, He founded it; He did not create it to be empty, but formed it to be inhabited—He says: I am the Lord, and there is no other. I have not spoken in secret, from somewhere in a land of darkness; I have not said to Jacob's descendents, 'Seek me in vain'. I, the Lord, speak the truth; I declare what is right." (Is. 45:18-19)

"Think of three things—whence you came, where you are going and to whom you must account."
—Benjamin Franklin

Chapter 14

Biblical History: The Age of the Earth and the Flood

Most studied churchmen are aware that the earliest biblical writings began during the time of Moses following the Exodus (1445-1440 B.C.). The age of Abraham, Isaac, Jacob and his 12 sons, including Joseph who was sold into Egyptian slavery by his brothers, extended over a period of 268 years (1708 B.C.).

The fall of Adam is dated approximately 1,600 years prior to Noah's flood (2200-2400 B.C.) which would date Adam's fall somewhere around 4000-3800 B.C.

The Bible does not record how long Adam and Eve remained in the garden; however, most Bible students believe that biblical records suggest approximately 6,100 years of human history. The earliest discovered non-biblical writings (on clay tablets) are dated around 3500-3000 B.C., well within the range of biblical history.

Prior to the writings attributed to Moses, the Hebrew people practiced what was common for other communities—oral tradition, in which stories were committed to memory and passed down from generation to generation. The stories of Adam and Eve, the tower of Babel, the flood and the later generations of Noah would fall into that category. The growth of a nation seeded by Abraham through the sons of Jacob and the Hebrew exodus from Egypt were also first recorded in the memory of what would one day become the nation of Israel.

The Age of the Earth

Can one believe that the earth is only 6,100 years old? The truth is that no verifiable writings, structures, tools, graves, meteorology or cave drawing exist that can be second-source verified beyond that point. Yes, some have indicated that radiometric or radio-carbon dating have both suggested dates beyond 6,100 years; however, those are all contested and cannot be relied upon as proof.

The Case for a Young Earth

1. **The Population Factor:**

   Today's population is approximately six and a half billion people. In 1985 the population was five billion, and in 1962 it was three billion. From the time of Jesus to the year 1810 the population increased from 250 million people to 1 billion. If one were to project on a graph back to the biblical year of the flood (2400 B.C.) based on known mathematical progressions of population, which take into consideration natural and expected catastrophes, that graph would clearly represent a starting point of two (or, as scripture states, four couples).

   Even if the earth experienced a devastating ice age 30,000 years ago and only one couple survived, our population over the next 30,000 years would be astronomical. Evolution teaches that man in one stage of evolvement or another has existed from one to six million years. Consider what that population number would be unless perhaps the earth has experienced an ice age and flood combo every 4,000 years. If evolution and an ancient earth are true and man arrived when evolution said he arrived, there would be 150,000 people per square inch of the earth's surface and there would be fossilized or frozen evidence to support it. One scientist suggested that based on present variables, if the world began with one couple 41,000 years ago, the present world population would be 2×10^{89} powers.

2. **The Moving Moon:**

   Astronomers agree that the moon is moving away from the earth at a rate of two to three inches per year. Most people realize that the ocean's tides are controlled by the gravitational pull of the moon as it revolves around the earth. The closer the moon is to the earth, the higher the tides. This would have been no problem 6,100 years ago; but one and a quarter billion years ago the moon would have floated just above the earth's surface. As the moon slowly moved away, the earth would experience constant and massive flooding, leaving very little time for Darwin's theory to evolve.

3. **The Earth's Magnetic Field:**

   Science agrees that the earth's magnetic field is getting weaker. Most scientists suggest something in the range of 10 percent over the past 150 years. Not only does this create major problems for carbon and radiometric dating, it also creates problems for an ancient earth. Based on present magnetic decreases the earth could not have existed more than 25,000 years and probably has an extended life-span of another 10,000 years.

But, give science its due. There is no way to determine whether the same rate of decrease existed 25,000 or more years ago, which is exactly what creationists have been saying all along and why carbon dating is only accurate up to 2,000 years according to many of those who do carbon dating.

Evolutionary scientists suggest that this decrease in magnetism is only a cycle and that it will one day reverse itself, though they don't offer any scientific suggestions as to how that might be possible.

Do you not wonder why some of those same scientists don't suggest that global warming or climate change might also be cyclical?

Though science has only calculated the decrease in the earth magnetic field for the past 150 years, it has not observed any slowing down or variation in that decline.

4. **The Earth's Rotation:**

The earth is spinning counter-clockwise at a rate of 465 miles per hour at the equator. Astronomers have noted that the earth's rotation time is decreasing at a rate of 1.7 seconds per year, requiring them to add a leap second from time to time in order that the earth's rotation will stay aligned with the international atomic clock.

This is no problem if the earth is 6,100 years old; however, if the earth is a billion years old, then winds on the earth would have been over 5,000 miles per hour.

Some evolutionists have attempted to solve this problem by suggesting that the earth must have rotated at a much slower rate millions or billions of years ago. As always, whenever verifiable scientific observation conflicts with the theory of evolution an exception is created in order to accommodate evolution, an exception that cannot be proven or disproven.

5. **The Sahara Desert**:

    The largest desert in the world grows larger every year as determined by the lack of moisture in the prevailing wind patterns. Based on the present size and rate of growth, approximately four miles every year, the Sahara Desert is estimated to be only 4,000 years old. Does that mean that winds were different before that time on earth or was there a catastrophic flood 4,000 years ago, after which deserts began to form?

6. **Check the Oil:**

    The earth is able to contain oil and gas pressure beneath the earth's surface at 20,000 psi. Science agrees that pressure build-up over more than 10,000 years would cause oil to burst through the earth's rock. Why do we still have oil under pressure beneath the earth's surface if the earth is billions of years old? Evolution has also told us that it takes millions of years to produce oil from decaying animals, though we have more recently learned that man can produce oil in 30 minutes from sewage and animal guts.

7. **Ice Rings and Tree Rings:**

The age of the earth could hardly be determined by ice or tree rings; however, if it were possible, science could only confirm an age of 4,300 to 4,600 years based on the earth's oldest tree (the bristlecone pine tree of Southern California) and ice cores from Antarctica.

8. **Niagara Falls:**

As a result of constant erosion, Niagara Falls is moving south at a rate of 4.7 feet per year. The Niagara Gorge is only 7.5 miles long. In the past 4,400 years it has eroded half that distance. If the earth is millions to billions of years old, why has Niagara Falls not eroded back to Lake Erie?

9. **Salty Water:**

Based on a constant erosion rate of soil minerals, the ocean is getting saltier at an estimated rate of 457 million tons per year. Today the ocean is 3.6 percent salt. When it rains 30 percent of that water goes into the ocean, the rest into the ground, which in turn continues to deposit salt minerals into the ocean by erosion. Evaporation only absorbs water not minerals though approximately 122 million tons leave the oceans by various other methods. Some scientists suggest that the present content of salt in the ocean would have been reached in 5,000 years. If only one tenth of the projected number of years that life is believed to have evolved on earth were factual, the oceans would be saturated with salt.

10. **A Flat Earth:**

Based on the present rate of erosion, scientists suggest that the earth will be flat in 14 million years. If the earth is billions of years old, why is the earth not flat? In addition, how can evolutionists explain why there are fossils discovered in sedimentary rock layers dated multimillions of years older than the present erosion rate would have allowed them to be found?

11. **Ancient Writings:**

The oldest reliable historical record is less than 6,000 years old. The Chinese calendar is based on the year of a flood—an estimated 4,700 years ago. The oldest writings discovered on clay tablets are estimated to be less than 5,000 years old.

12. **Comets:**

Comets lose mass (material) and are only able to last (maintain mass) approximately 10,000 years. Why do we still have comets? Some evolutional scientists suggest the possibility of an "Oort Cloud" that continues to produce comets, though there is no way to prove it exists. This theoretical cloud is suggested to be 50,000 astronomical years (AUs) away from earth while Pluto is only 30 AUs away. Based on current astrological observations, the existence of comets is not possible if the earth is over 10,000 years old.

13. **The Muddy Mississippi:**

The Mississippi River deposits 80,000 tons of sediment every hour in the New Orleans delta, which continues to grow larger and larger. Based on its hourly deposits and the present amount of sediment in the delta, it is estimated that the delta is 30,000 years old. If the earth is billions of years old why is the Gulf of Mexico not full of mud? Creationists suggest that the delta's present collection of mud is due to a massive flood 4,500 years ago.

14. **The Great Barrier Reef:**

The Great Barrier Reef off the coast of Australia is the largest coral reef in the world. Following World War II, environmentalists observed it for about 20 years in order to determine its rate of growth. They calculated that the reef was only 4,200 years old. Why is the reef not much larger if the earth is millions or billions of years old, or did the Great Barrier Reef begin growing following a worldwide flood?

15. **A Declining Sun:**

Though the sun tends to swell and decline, the observed pattern of the sun is that it is in decline at a rate of .1 percent every century or five feet in diameter every hour. The present mass of the sun is 1.8 octillion tons (330,000 times greater than the earth), and it is certainly able to sustain present earth conditions for many centuries to come. Based on the present rate of decline and by extrapolating its mass content over the past 258 million years, the sun would have at that time been touching the earth's surface.

16. **Helium Diffusion:**

    As elements decay they produce helium which escapes from rocks into the atmosphere. Based on the present helium content in the atmosphere, some scientists suggest that the earth could not be older than two million years.

When is Enough, Enough?

Evolutionists face a difficult task. They must explain all the preceding evidence for a young earth, and at the same time construct a theory of creation consistent with nature, science and logic—all of which, every step of the way, must be true for evolution to be true.

Consider a few more contradictions before I remind you of the one thing that enables evolution to continue as a theory:

1. Before landing on the moon astronauts were worried that their landing craft would sink into deep layers of dust accumulated over billions of years. They were relieved to discover that there was only a half-inch layer of cosmic dust on the moon, which indicated the moon has not been collecting dust for billions of years.
2. Though meteorites are very rare, none have been discovered below the earth's top layer, indicating the earth has not been exposed to meteors for millions or billions of years.

3. The present rate of star cluster expansion denies a billion-year old universe.
4. Jupiter and Saturn are losing heat twice as fast as they are able to gain it from the sun. They cannot be billions of years old, as they would not exist.
5. The present volume of volcanic lava divided by its rate of efflux only allows for a few million years since creation.
6. Helium 4 is forming faster in the atmosphere than a billion-year old planet would allow. Based on the present helium 4 content, the earth could not be older than 175,000 years.
7. The small amount of ocean sediment indicates a young earth.
8. The largest stalactite in the world could have easily formed in 4,500 years.
9. Ice formation at the poles indicates a 5,000-year old earth.

Only one reality exists that allows evolution to continue as a theory, and it will continue to do so. There will always be those who deny the existence of God and refuse to allow for a divine creator. Yes, many God-believers have been persuaded to believe that evolution was a scientifically proven fact; they are not the life-blood of evolution's survival. Do not be deceived; evolution is the religion of the godless.

Just as the theory of evolution requires millions and billions of ages to be taken seriously, the biblical account of creation must also be chronologically realistic and bound together with a logical progression of historical events as they relate to creation, including the Fall of man and the catastrophic Noah's flood.

God lives outside the realm of time, space and substance and certainly could have used whatever time span He desired to create the universe. He also could have begun life on earth with a single designed cell (ancestor) had He chosen to do so, but the evidence and Scripture point in a different direction.

The theory of evolution and those who are so determined to prove the Genesis six-day account a myth and, thereby, deny the existence of God, have fueled their argument with assumptions based on what must be possible if evolution is true, not what can be observed and proven. They say the layers of the earth appear to be the result of millions of years of erosion, though the evidence points otherwise. They say that fossils in sedimentary rock appear to demonstrate an evolutionary progression of lower to higher life-forms, though the evidence indicates otherwise. They say various dating techniques reveal a very old earth, though few scientists are willing to accept their reliability without a verifiable second source. They say life mysteriously started from non-life, though that proposal would invalidate everything science now understands about chemistry and biology. They say that all species on earth have evolved from a single ancestor, though all that evolutionists can point to are DNA similarities of kind and the perchance possibility of complex mutations over eons of years.

Noah's Flood and Other Flood Stories

The physical evidence for a catastrophic flood is so overwhelming that some evolutionists have begun to suggest a massive monsoon or several massive floods must have occurred in past years.

I have not counted them, though I have read some of the hundreds of flood stories collected from cultures all over the world.

The interesting thing about all these flood stories is that the vast majority include the same central components as Noah's flood: (1) a warning to the people (2) construction of a boat (3) the storage of animals (4) family included in the boat and (5) the later release of birds. Did the Bible copy another source or was this event so monumental that it remained in the memory of people as they were later scattered around the world?

Divine judgment and the rescue of some in a vessel is an almost universal theme. Some of the more famous flood stories include the Babylonian Gilgamesh Epic; recorded on clay tablets, it is perhaps the oldest written record of a catastrophic flood. The Egyptian Book of the Dead, China's Hih King Flood, several flood stories from India and even the American Eskimo and several American Indian tribes had their own version of a flood story.

It would be difficult to believe that so many cultures could have conspired to create a myth or that so many stories about a catastrophic flood could have evolved completely independent of a common event.

Would a Catastrophic Flood Produce Layering?

Were the layers of the earth created as a result of a massive flood or were they created over long periods of time by erosion? Rather than test your power not to fall asleep during this next section, allow me to quickly summarize some of the material already covered in a previous chapter.

1. Would slow erosion over multimillions of years create uniform layering of the earth or, as I believe, no layering at all? As rock is formed by heat and pressure, sediment being constantly washed (eroded) from higher mountains to lower plains or valleys would be all, over time, solidified into mostly metamorphic rock. There would be no layers with loose sedimentary rock between them unless those layers were more recently formed and formed quickly, perhaps as a result of massive volcanic dams being created and destroyed over and over again by unrelenting worldwide seismic activity and flood waters.

   What we observe in the erratic layering of the earth is exactly what would have been expected by massive flooding, not slow erosion over long periods of time. The layers are wavy, inconsistent in number and definitely not uniform.

   Evolutionists who attempt to paint the picture of an ancient earth do not remain true to their own theory. What should have been created over 4.5 billion years is a consistent record of uniform erosion accompanied by temporary regional and natural occurrences like volcanoes and earthquakes, etc; however, what we see in reality is inconsistent layering in size and number, not uniformity.

2. What was presented as a uniform progression of smaller and less complex life-forms in lower strata and larger and more complex life-forms in higher strata based on 19th-century science and fossil discoveries has been completely overturned by recent new discoveries. The orderly progression of fossils over

millions of years in a geologic column is an absolute must if evolution is to be believed; however, more recent discoveries have yielded a total mixing of all life-forms throughout the fossil record. That, of course, is what led to the more recent evolutionary theory "the Cambrian Explosion", which in long-term evolutional ages is impossible.

I hope readers realize how damaging more recent fossil discoveries are to the theory of evolution. With no way to date fossils uniformly through what was originally thought to be ancient layering of the earth, evolution is left holding the myth, not six-day creationists.

These findings confirm that all organisms were living together and, as the Bible records, many were destroyed together. One thing that should be mentioned is that carbon-based and formerly living organisms do not require thousands or millions of years to fossilize. Organic matter exposed to air will decay; however, that same matter shielded from air and external exposure will often mineralize. This process can sometimes take only a few years and does not require thousands or millions of years to occur.

Following the 1980 eruption of Mt. Saint Helens in Washington State, approximately 20,000 trees were deposited in Spirit Lake, located north of Mt. Saint Helens; they are now beginning to fossilize.

3. The fossil record reveals what might naturally occur in a worldwide catastrophic flood including trees, now fossilized, propped vertically and upside down be-

tween two and sometimes three strata that were, according to evolutionists, supposedly separated by millions of years, as opposed to days or weeks in a catastrophic flood. These trees can be observed in many locations around the world and their presence should be evidence enough to disprove a long-term geologic column and suggest, rather, massive flooding over a short period of time.

In addition, discoveries continue to be made all around the world of artifacts in lower layers of Lyell's column that only a massive flood could explain. For example, a clay doll was found in rock evolutionary-dated 12 million years ago. Human tools were found in a stratum dated 55 million years ago. According to one source, a lady in Illinois found a 10-inch gold chain inside a lump of coal.

There have been thousands of frozen leaves found in Antarctica where there are no trees. A decorative bell, an iron pot and the sole of a shoe have been found in coal supposedly deposited multimillions of years ago (Powder River Basin Mine in Montana).

Emotional Agenda

Creationists as well as evolutionists can only offer observation, logic and evidence. Neither position can be scientifically proven because neither belief can be scientifically recreated or empirically demonstrated. Once an emotional agenda is removed from one's belief, the miracle of God's creation should speak for itself.

One tool of debate exists that is worse than emotions—and that is fraud. This study has attempted to present the facts and findings from many verifiable sources and all are subject to your own research. It has not attempted to proof text every finding or give credit to the research, expertise and discoveries of all the many real experts in this field, though, once again, all are subject to your own discovery.

Emotion or pride has driven many in the study of evolution to "cook the books." Their firm disbelief in a God has convinced them that only evolution is possible; therefore, the end often justifies the means when it comes to evidence. For a Christian the end should never justify the means. God desires to commune with only those who desire to commune with Him with a righteous heart. "Draw near to God and He will draw near to you" (Jas. 4:8). Those who truly seek Him will not be fooled by the false claims of evolution or any other anti-God theory. I am not talking about questions and doubts that are a part of man's human spirit. I refer to those spiritual beliefs, values and virtues that God places within the believer by way of the new spirit He gives to all citizens of His Kingdom.

The Waters of Judgment

Though I refer to the biblical flood as Noah's flood, it is actually God's flood in which Noah and his family received deliverance from God's judgment. Judgment is a concept few people are willing to entertain, as they prefer to envision God as only a grand old man full of love and compassion.

Why would God completely destroy what may have been a population of millions of human beings? Should God not take some responsibility for creating man with the freedom to sin and disobey Him?

The Psalmist wrote, "Surely your wrath against men brings you praise, and the survivors of your wrath are restrained" (Ps. 76:10). Every parent knows or has learned the hard way that judgment is a required ingredient in parental love. It was by the hand and will of God that man was given dominion over the earth; however, man was not given dominion over right and wrong. The Prophet Habakkuk wrote, "But the Lord is in His holy temple; let all the earth be silent before Him" (Hab. 2:20).

In today's world of human "rights", where every person is so concerned about his right to do this or to do that, we humans need to be reminded that God determines the "rights" He chooses to grant—not man.

Man is, at times, blinded by the beauty of creation and forgets the will of the creator. In their quest for earthly pleasure and dominance, men rebelled against the giver of life and the righteous joy that God intended to provide to mankind. Silence ("hacah") before God is a hushed stillness in recognition of God's supreme rule as the giver of life and continued breath. Judgment against anything or anyone that seeks to destroy God's will is the inevitable consequence of rebellion against God.

Strange Findings

Evidences of God's creation and His continued involvement in this earth plainly reveal God's handiwork. Noah's flood is high on that list. Prior to this catastrophic event fairly clear evidence exists that the earth's atmosphere was far more oxygen-rich than it is today. The earth was shielded from the destructive gamma and ultraviolet rays of the sun by a firmament above the earth. The Bible refers to this firmament as the waters above the earth's waters (see Gen. 1:17).

Plants and animals lived longer and grew larger during that time (approximately 1,600 years prior to the flood). The Bible talks of men living almost a thousand years and also "Nephilim (giants) were on the earth in those days" (Gen. 6:4).

Perhaps you thought this too was mythical or dismissed it as hyperbole (intentional exaggeration). Consider some recent discoveries, however, that reinforce biblical truth:

1. In 1950 a fossilized human femur was discovered in Turkey that indicated a man 14- to 16-feet tall.
2. Insects breathe through their skin and they also grew larger. A fossilized dragonfly with a 50-inch wing span, an 18-inch cockroach, an 8.5-foot-long centipede, a two-foot-long grasshopper and a tarantula with a three-foot leg span have all been discovered in the fossil record.
3. Fossil evidence has also indicated that animals grew larger, including a donkey in Texas that was nine feet tall at the shoulders. A buffalo with a 12-foot horn span, an elk with a 12-foot antler span, 10-foot-tall kangaroos, giant lions and wombats the size of a small car, a 1,000-pound goose as tall as an elephant and eight-foot-long beavers found in Ohio and Wisconsin have all been discovered.
4. Plants and trees grew larger, including the discovery of 60-foot cattails in sedimentary rock.
5. Other recent discoveries include birds 13 feet tall, 11.5-foot-wide oysters found in Peru two miles above sea level, a shark's tooth indicating a 60-foot-shark as well as other large fish and turtle fossils.

6. Large dinosaurs could be considered in a category by themselves as a considerable amount of fossilized and non-fossilized bones and partial skeletons have been discovered all over the world. If one accepts the premise that at one time before an ice age or a massive flood that the world's atmosphere was vastly different and that animals as well as plants on which the animals fed grew much larger, as many discoveries have indicated, then it is not hard to understand how some dinosaurs who lived long lives also grew very large. The brachiosaurus (34 tons and 46 feet tall) had nostrils the size of a modern horse and would have suffocated in today's atmosphere.

Some scientists believe there were as many as 600-1,000 types of dinosaurs, though creationist Ken Ham suggests probably only 50 different "kinds" existed.

The conflict between creationists and evolution is not with the existence of dinosaurs, but rather, how long ago they lived. The Bible speaks of dragons ("tanniyn") as serpents or large land and/or sea creatures. In Jeremiah 51:34, "He has swallowed me up like a monster", the meaning is clearly a creature large enough to swallow a man. Similarly, Ezekiel 29:3 refers to Pharaoh, king of Egypt, and compares him to a great sea monster.

Isaiah 27:1 speaks of God destroying leviathan, representing the enemies of Israel. Many cultures, including the Hebrews, spoke of monsters, dragons and large land and sea creatures.

The Book of Job is perhaps the most unique body of writing in the entire Old Testament. Many scholars believe that Job lived prior to the time of Abraham, during that period of expansion following the flood. If that is true, then Job cannot be considered a Hebrew, though Job definitely demonstrates a belief in a creator God, which may have been passed down from Noah himself.

In chapter 40 and 41 Job speaks of behemoth and leviathan as real creatures created by God. "Look at the behemoth, which I made along with you and which feeds on grass like an ox . . . his tail sways like a cedar . . . his bones are tubes of bronze . . . he ranks first among the works of God" (see Job 40:15-19). In the very next chapter Job speaks of a different sea creature: "Can you pull in a leviathan with a fishhook? His snorting throws out flashes of light, his eyes are like the rays of dawn. Fire brands stream from his mouth . . . smoke pours from his nostrils . . . when he rises up the mighty are terrified . . . iron, he treats like straw and bronze like rotten wood" (see Job 41:1-27).

Were these animals living during the time of Job? Were they remembered by the descendants of Noah following the Flood? Remember, Noah was still alive when Abraham was born, though probably not at the time Abram was called out by God.

In the Sumerian Epic, Gilgamesh is said to have slain a huge dragon (2000 B.C.). Alexander the Great discovered the Indians worshipped large hissing snakes (330 B.C.).

Now, before you call me crazy, check out for yourself the many reports from around the world of dinosaur-type creatures still roaming the earth today. Two separate sightings were made by over 1,000 people in China of a dinosaur-type creature. A Canadian professor told a meeting of zoologists about the reported 300 sightings of a Cadborosaurus (a sea creature with a horse-like head and hind flippers) around the coast of British Columbia. The remains of a young "caddy" were found in the stomach of a whale and pictured in the newspaper. Non-fossilized dinosaur bones are found quite often, including those said to come from a horned dinosaur, a duckbill dinosaur and others.

Along with many pictures of dinosaur-type creatures found in cave drawings, the American Indians told stories of a winged creature resembling a Pterosaur (a winged lizard said to have died 65 million years ago). Several other sightings of this creature were reported during the 20th century. In 1990 a group of scientists from the University of Montana found partially fossilized T-Rex bones with hemoglobin still present in the red blood cells. Also in the 1990s, explorers in Nepal found elephants with similar characteristics as mammoths.

Stories of dinosaur-type creatures, including Scotland's Loch Ness monster, sworn to by hundreds of people, have also come out of Africa, Russia and, most recently, New Guinea.

Now, one must readily admit that there have also been stories of alien beings from other planets coming from many locations as well; however, you can research and judge for yourself.

Did Dinosaurs Live with Humans?

While fossilized human and dinosaur bones have never been found buried together, recent discoveries completely overlooked or summarily dismissed by evolutionists would indicate that, in fact, they did live together. Cave drawings in Peru have depicted men and dinosaurs together. Human hand and footprints have been discovered together or in the same location with dinosaur prints all over the world.

One such discovery was made in Glen Rose, TX, on the Paluxy River where human footprints 15- to 20-inches long are clearly present beside dinosaur prints found in limestone. Some human prints were also found inside dinosaur footprints. One set of human footprints indicated a human stride of six to seven feet.

I must qualify these findings as evolutionists sometimes will insist that these findings are not human footprints. It was said that a well-meaning creationist attempted to brush or dust these footprints before taking pictures, thereby destroying their integrity. Others, based on their inspection, believed these prints to be smaller dinosaurs.

The discoveries in Glen Rose, TX, began in 1902, but it was not until 1980 that television and local newspapers began reporting on them. Excavation can only be done for about two weeks every year as the Paluxy River is too high at other times. Several other accredited scientists have testified to the authenticity of both sets of prints.

In a mountainous region of Turkmenistan, Russia, several thousand dinosaur tracks were discovered in what is labeled the Jurassic layer of rock. In addition to the 65 distinct types of dinosaur tracks, there were also discovered human or human-like footprints on the same plateau. This particular plateau is said to contain the largest fossilized dinosaur track-way in the world.

According to one expedition leader's evolutionary timetable, he describes all the fossilized prints as 150 million years old. This, he said, "is in direct contradiction to the evolutionary theory."

In the year 2000, Dr. Dennis Swift of the Dinosaur Institute was invited to inspect the area and he said that "there were six human prints on one slab", he observed. All were located near dinosaur tracks in the same area. Several other scientists have identified these human tracks as authentic.

Before leaving this topic, allow me to remind you of one additional inconsistency with a 60- to 150-million-year old dinosaur evolution. From time to time non-fossilized dinosaur bones are discovered in tar pits. The question that is being asked is how is it possible for these bones to still contain blood cells including the presence of hemoglobin (protein matter) in the red blood corpuscles? These blood cells should have decayed within several thousand years, and yet, they remain present. Think carefully. There are only three questionable sources of evidence for believing that dinosaurs lived 60 million years ago.

1. The jumbled Geologic Column,
2. Radiometric dating based on a rock or fossil's location in the geologic column and . . .
3. Imagination—that's it.

Giants

Giant human skeletons and bones have been found throughout the United States, Sweden and Mexico. One of the more notable discoveries was a fossilized giant unearthed in Ireland that measured over 12-feet tall and weighed in at two tons. It was displayed in both Liverpool and Manchester, England, and was pictured in several newspaper articles leaning up against and towering over a train boxcar.

A 35-pound ax head was discovered that indicated only a very large man could have swung it. Many skeletons, partial skeletons and bones have been discovered that attest to the presence of men nine- to 12-feet tall living on earth.

In Numbers 13:32-33 it is recounted how men sent out to explore Canaan, the Promised Land, reported finding giants (Nephilim). "We seemed like grasshoppers in our own eyes, and we looked the same to them." In 1 Samuel 17:4, we read that David the shepherd boy defeated Goliath, a giant over nine-feet-tall, and in 1 Chronicles 20:4 we read that other giant descendants of Rapha were also killed by the Israelites. One other descendent of Rapha is mentioned in 2 Samuel 21:15-17, in which it says one of David's soldiers saves him from being killed by a giant.

The significance of these findings around the world serves to validate scripture and at the same time question an evolutionary theory of continuously descending larger and stronger life-forms.

Dragons, giants and a worldwide flood have often been relegated to the realm of fantasy in today's scientific (empirical) world; however, even scientists are learning that what might appear supernatural today may have been the natural world of years gone by.

My son introduced me to a book in which I discovered that early map-makers would identify lands that up until that time had remained unexplored as "Here be Dragons."

> "Let the wise listen and add to their learning, and let the discerning get guidance—for understanding proverbs and parables, the sayings and riddles of the

wise. The fear of the Lord is the beginning of knowledge, but fools despise wisdom and discipline. Listen, my son, to your father's instruction . . ." (Pr. 1:5-8)

Many Christians find themselves embarrassed to admit they believe in a six-day creation. Why is that? Why do some try to accommodate invalidated evolutional beliefs with biblical beliefs? Many people live by the axiom of "we don't talk about religion and politics." Well, based on what's happening in the world today, perhaps we should start. The next chapter illustrates how the world seeks to silence those who trust in God and God's Word.

Chapter 15

Arguments of Last Resort

As I have read through the countless articles and books that criticize those who believe in the biblical account of creation, they all come back to the same general arguments:

1. "Creationists just show how ignorant they are about evolution and science in general."
2. "Just because someone can punch holes in some parts of the evolutionary theory, that doesn't mean that, by default, God or creation by God should be taken seriously."
3. "Religion has always stood in the way of science. If it were left up to religion, mankind would never have advanced its knowledge of the universe."
4. "God is such an improbable being that it should require extraordinary proof before being seriously considered as truth. If God exists, then He has some serious explaining to do because His creation is a real mess."

All four of these brilliant arguments, as far as I am able to understand them in my state of ignorance, can be boiled down to one simple statement—there is no god and only the ignorant would believe there is. To the devout evolutionist that is exactly what the real debate is all about—the existence or nonexistence of God. It is most often the angry evolutionist who creates the heated debate—angry at God and angry at those who believe in God. It is also true that the angry person is the one who gets the attention. Be that as it may, angry people can also be somewhat careless with the evidence and logic.

Men of Science—Men of Faith

I dare say our evolutionary friends would not label Sir Isaac Newton, Galileo, Nicolaus Copernicus, Francis Bacon, Rene Descartes or Louis Pasteur, all of which believed in God, ignorant—nor would Albert Einstein, Albert Schweitzer, Martin Luther King or T.S. Eliot fall into that category. Physicists Max Born, Emilio Segre, John Cockcroft and Bertram Brockhouse, among many other Nobel Laureate winners, all believed in God.

In 1952, Nobel Laureate winner in medicine and physiology Alexis Carrel said, "Jesus knows our world. He does not disdain us like the god of Aristotle. We can speak to Him and He answers us . . . He is God and transcends all things."

In 1956, Nobel Laureate winner in physics, Arthur Compton said, "I see Him [Jesus] as the Everest among the world's many high mountains."

In 1982, Nobel Laureate winner in physics Robert Millikan said, "Jesus preached it [service] as duty—for the sake of world salvation. Science preaches it as a duty for the sake of world progress." Millikan believed that the "combination of science and religion provides the sole basis for rational thought."

Winner of the Presidential Medal of Freedom and Nobel Prize for Literature, poet and playwright T.S. Eliot said, "Christ is the still point of a turning world."

Sir Isaac Newton said, "The most beautiful system of the sun, planets and comets could only proceed from the council and dominion of an intelligent and powerful Being."

Galileo Galilei (1564-1642), physicist, mathematician, astronomer and philosopher, said, "When I reflect on so many profoundly marvelous things that persons have grasped, sought and done, I recognize even more clearly that human intelligence is a work of God, and one of the most excellent."

According to Dr. Russell Humphrey, physicist, he estimates, "There are around 10,000 practicing professional scientists in the U.S. alone who openly believe in a six-day recent creation." Biologist Wayne Friar, *"Case for Creation"*, suggests another 25,000 United States scientists "reject the evolutionary doctrine that all living organisms are related."

And the list of "ignorant" people, all of whom believe in God, goes on and on.

Punching Holes

It is not my intent, nor should I be so disingenuous as to suggest, that I want to "punch holes" in the evolutionary theory. I intend to knock it through the ropes and out of the ring of debate. Anything short of that will be doctored up by the ringside corners. Too many loopholes, anomalies, exceptions or things to be discovered later exist with which to compete. No, to some, God will not be left standing, but only because that is their final rebellion.

For some, evolution is the crutch that enables them to "intellectually" deny God. But worse than that, evolution has become the tool whereby those who reject God can influence others to join their cause. It is not only God-believers who join congregations; it is also disbelievers who seek their own following and confirmation.

The Improbability of God

Many of those who choose to reject God use as their rationale the existence of evil in the world, suffering from natural or unnatural catastrophes and what they view as God's seeming unwillingness to address these problems. In fact, God has addressed these concerns, because they were His concerns first.

How many of those who complain about the problems of the world would prefer a life without free choice? Personal responsibility and free choice are Judeo-Christian absolute principles, but to the godless it is all a matter of chance. Granted, many people are born into deep poverty and dangerous circumstances; however, even among extremely disadvantaged people, how many would intentionally give up personal freedom in order to enslave themselves to the whims of another? Perhaps, only as long as it took for them to be released from those disadvantaged circumstances, then they too would seek their freedom of choice.

At some point choices are made that not only affect the individual but also affect generations to come. Some are beneficial to later generations and other choices are not. To build a community at the foot of Mt. Saint Helens or along a turbulent sea coast is a choice not dictated by God.

God has provided all the essentials necessary to make this a beautiful planet. He has born into man the knowledge of right and wrong (moral awareness); and yet, man continues to abuse the planet and abuse one another. Who can say but that God did not provide the one who would have one day discovered the cure for cancer or a formula for ColdFusion, but the self-consumed sin of a woman's free choice destroyed that child while still in the womb? Life is a series of choices because God desired to make it so.

When God created man He gave him one command, and man chose to disobey. It was man's choice; and yet, it had consequences. God compels man to make good (righteous) choices; He does not force man. God loves and desires love; however, He does not force love for Himself or for others.

If it is as some have suggested that the world is in a real mess, is it God who should answer for that? How much further must God go to redeem that mess than He has already gone? Must He continue to enter this world, suffer and die over and over again until man chooses to clean up the mess man's choices have created?

You see, either free choice exists or it does not. Either the complexities of nature exist, or nature ceases to exist. God is creator not dictator, He is sustainer not puppeteer and He is deliverer, if one elects to accept His offer of deliverance.

Do you not recognize that in one's humanness enough is never enough? Stop the plane from crashing to the ground by eliminating gravity. End death and allow people to live until their body simply decays away. Give me one or two more years to live and I'll be content. How can anyone look at this world and say the mess belongs to God?

You who say you don't believe in God because you are angry about suffering—do you not think that believers have also been angry with God for their own suffering? Anger is an emotion and believers are, perhaps, the most emotional of all people.

What separates a believer in God from a non-believer is most often trust that seeks to overcome emotional anger. Trust in God's ultimate deliverance (salvation) and trust in the eternal nature of God's love and mercy.

If anyone had just cause to be angry with God, it was the prophet Jeremiah. He was hated as a traitor by his countrymen. He was told by God to prophesy the destruction and defeat of Judah and because of that he was thrown in prison and later into a cistern to die.

Jeremiah said, "Lord, you deceived me . . . you overpowered me . . . I am ridiculed all day long; everyone mocks me . . . the word of the Lord has brought me insult and reproach." Jeremiah continues, "All my friends are waiting for me to slip, saying 'perhaps he will be deceived; then we will prevail over him and take our revenge on him.'"

In spite of that, Jeremiah remains faithful. "But the Lord is with me like a mighty warrior, so my persecutors will stumble and not prevail . . . for to you I have committed my cause" (I have placed my trust). (See Jer. 20:7-12).

While I too at times have been angry at God over personal suffering and the mess of the world, I have been far angrier at myself for my lack of trust. God's word "teaches us to say 'no' to ungodliness and worldly passions, and to live self controlled, upright and godly lives in this present age, while we wait for the blessed hope—the glorious appearing of our great God and Savior, Jesus Christ, who gave Himself for us to redeem us from all wickedness and to purify for Himself a people that are His very own, eager to do what is good" (Titus 2:12-14). Has it not been man's own ungodliness that has brought about much of the suffering?

You see, only God is capable of redeeming this world. God's plan is focused. It does not require long periods of idle time. It does not require the death of millions of transitional life-forms waiting for the fittest to survive.

With God, the end was created at the beginning; and in the end all things will not only be revealed, they will also be made new. When God called His creation "good" He had full knowledge of both the beginning and the end.

"Then I saw a new heaven and a new earth, for the first heaven and the first earth had passed away, and there was no longer any sea [separation between peoples of the new earth]. I saw the Holy City, the New Jerusalem, coming down out of heaven from God, prepared as a bride beautifully dressed for her husband [Christ]. And I heard a loud voice from the throne saying; 'Now the dwelling of God is with men, and He will live with them. They will be His people, and God Himself will be with them and be their God. He will wipe away every tear from their eyes. There will be no more death or mourning or crying or pain, for the old order of things has passed away. He who was seated on the throne said; 'I am making everything new'" (Rev. 21:1-5).

Ron Edwards, a friend who is far more science-oriented than I, believes that when John speaks of the first heaven and the first earth passing away that Christ will literally release nuclear control over the atom (matter). Just as John's Gospel speaks of all things being made (energy) in Christ (the Light), all neutrons, protons and electrons will be released by Christ and matter will vanish.

There is hope, however, even for those who anguish over this present sinful earth. Faith, trust and hope are what sustain the people of God.

"For in this hope we are saved. But hope that is seen is no hope at all. Who hopes for what he already has? But if we hope for what we do not yet have, we wait for it patiently." (Rom. 8:24-25)

Chapter 16

The Cosmological Search

> "Science without religion is lame; religion without science is blind"
>
> Albert Einstein

One of the most understandable presentations of recent cosmological findings is found in a book entitled, "I Don't Have Enough Faith to be an Atheist", by Dr. Norman L. Geisler and Frank Turek, both highly respected authors and teachers in the fields of theology, philosophy and science. They have explained in layman's terms several recent Nobel Prize discoveries that many scientists have labeled the most significant discoveries of the 20th century, if not, for all time.

Background

The Law of Causality is the under pinning of all scientific research. Science, itself, is the empirical search for causes. As far back as Aristotle, we were exposed to the theory of a universe in flux (motion). As motion required a mover, the first motion implied a first mover—a first cause.

The Cosmological Search is the quest for understanding the beginning of the universe. If the universe had a beginning, then the universe had a cause according to the Law of Cause and Effect. The argument is framed as follows:

1. "Everything that had a beginning had a cause,
2. if the universe had a beginning,
3. ... the universe had a cause".

Francis Bacon said that "true knowledge is knowledge by causes." David Hume wrote, "I never asserted so absurd a proposition as that something could arise without a cause."

For centuries many outside the Judeo-Christian community believed that the universe (matter and substance) including our solar system was eternal—it had no beginning and, therefore, no cause. Early in the 20th century that theory began to come unraveled with Albert Einstein's Theory of Relativity.

Einstein had long held to the belief that the universe was eternal with no beginning; however, his calculations were destroying that belief. What ensued were two important discoveries that confirmed for Einstein what is now universally accepted, that General Relativity (the effects of gravity on motion) is true, the universe is growing, it is not static and it must have had a beginning.

The first experiment to contradict the eternal universe theory was conducted by the British cosmologist Arthur Eddington during a solar eclipse. To his dismay, Eddington wrote, Einstein's theory was correct and he added, "Philosophically, the notion of a beginning of the present order of nature is repugnant to me . . . I should like to find a loophole."

Later astronomers, including Edwin Hubble of the Hubble telescope, observed what was described as evidence of an expanding universe. Logic dictated that if the universe was expanding then the universe obviously had a beginning.

Einstein himself traveled to the California observatory where Hubble had discovered a "Red Shift" (also called the Doppler Effect), in which wave lengths increase from their source towards the less energetic red end of the electromagnetic spectrum, coming from every galaxy. The "red shift" was strong evidence that those galaxies were moving away from the earth.

Dr. Geisler calls this cosmological discovery "the beginning of the end for atheism." He frames his discussion around five major points (universal laws and recent discoveries). At first many, if not most, scientists were reluctant to accept the notion of a beginning that logically followed from an expanding universe. Their solution of extracting God from the equation as the first cause came in the form of "the Big Bang Theory." Theories are abundant, the devil is in the details (proof or, at least, observable evidence).

As it was originally proposed, the "Big Bang" was a theory that said the universe is constantly contracting into nothingness, which eventually explodes (Big Bang) into something (the universe). This formula was said to repeat itself over and over again approximately every 80 billion years (Bang, Bang, Bang, etc).

What now follows is Geisler and Turek's five-point explanation of the true nature of our universe.

1. **The Second Law of Thermodynamics**

   As we have earlier discussed the Laws of Thermodynamics, I only offer this quick reminder. The first Law of Thermodynamics states that the universe (even the expanding universe as we have now discovered) is a closed system. The total amount of energy is constant. As no new energy is entering our universe we are left with a finite amount of energy.

The second Law of Thermodynamics is the Law of Entropy (inertness). Energy used becomes energy unavailable. Fire consumes the log and no longer produces heat. The car runs out of gas and the car stops.

It is the Second Law of Thermodynamics that tells us that the universe had a beginning. Because we still have order (energy available), the universe (substance dependent upon energy) is not eternal. The universe had a beginning; if not, it would not exist. The fact that it does exist means that the universe had a beginning and as yet, has not reached equilibrium where no new energy is available.

2. **The Universe Is Expanding**

   For some time the Laws of Thermodynamics had been used by creationists to discount the claims of evolutionists; now they are being used to disclaim the many "Bangs" theory. Some evolutionists proposed an exception to the Laws of Thermodynamics that somehow allowed those laws to be suspended in order for the needed energy to produce the universe.

   With the discovery of an expanding universe, the Big Bang Theory took on a whole new light. It was not only discovered that the universe was expanding, it was also discovered that space, itself, was expanding. How could an ever-expanding universe be consistent with repeated Big Bangs?

Einstein had previously predicted an expanding universe, but it was not until scientific observation became the evidence necessary to demonstrate expansion of the universe. There was no time, space or substance before the first and only possible "Big Bang" (the beginning of time, space and substance).

British author and atheist Anthony Kenny wrote that if the Big Bang is proven then one "must believe that the matter of the universe came from nothing and by nothing." As with all those who reject the existence of the creator God, "nothing by nothing" is all they have to which to cling. Aristotle wrote that nothing is what rocks dream about.

3. **Radiation from the Big Bang**

Later, two Bell Lab scientists accidently discovered what is referred to as the afterglow from the Big Bang explosion. The afterglow is actually light and heat radiation that is observed in wave lengths, wave lengths they discovered that were coming in all directions. With their discovery, for which they received the Nobel Prize, the notion that the universe was an eternal reality was forever laid to rest. "The wave lengths are exactly what one would expect to observe as the pattern of light and heat produced by a great explosion", according to astronomer Robert Jastrow. But there's more . . .

4. **Great Galaxy Seeds**

Astronomers, having confirmed their new theory of an expanding universe as observed by light and heat radiation in wave lengths, now turned their attention to the possibility of observing ripples or slight temperature variations within the light and heat radiation. Many believed that in order for galaxies and planets within galaxies to form that the temperatures required to produce the necessary and exact gravitational balance between those solar bodies would in turn produce temperature ripples within the radiation field. These ripples would demonstrate how matter was allowed to congregate by gravitational attraction into solar bodies.

Their belief was confirmed when in 1989 NASA launched a satellite specifically designed to observe and measure heat ripples.

Astronomer Stephan Hawkins called this discovery "the most important discovery of the century, if not of all time." Others called it the "Holy Grail of Cosmology". When NASA's findings were announced in 1992, astronomer George Snoot said, "If you are religious, it's like looking at God."

What these ripples (great galaxy seeds) demonstrated was, as Snoot termed it, "the fingerprints of the maker" and irrefutable evidence of an expanding universe. What was as remarkable as the discovery of temperature ripples was the precision in which they occurred allowing just the right amount of matter to congregate into orbital planets and galaxies without colliding in on themselves (remember the "irreducibility of complexity").

Before concluding the Cosmological Search allow me to summarize the evidence as well as the reason these discoveries are so significant to God-believers.

For both a theist and an atheist, matter had to come from somewhere, because here it is. For the atheists, their early solution was to say that matter (the earth, moon and stars) were pre-existent, it had always been here. As later discoveries, including those we have now discussed, yielded evidence that the universe actually had a beginning, rather than attributing that beginning to God, they favored a new theory that indicated the universe was eight to 20 billion years old and the result of a Big Bang explosion of dust and chemical gases. This new theory, however, violated the same natural laws that had been violated by evolution's old theory. It was based solely on the notion of what must be true if there is no creator rather than scientific principles or cosmological discoveries.

In truth, the age of the universe is not the significant issue, nor does it answer the question as to whom or what began the universe. What it does prove, however, is that the universe has an age, which means it had a beginning. And that is significant.

The "Big Bang", originally adopted by evolutionists as a dismissal of a god of the universe based on their assertion of repeated "big bangs" every 80 billion years or so, is now embraced by God-believers as evidence of a one-time "big bang" approximately 6,000 years ago (some creationists believe approximately 6,000 to 10,000 years ago).

5. **Einstein's Theory of General Relativity**

"Then God said, 'Let there be light,' and there was light." (Gen. 1:3)

The nature of the universe and the biblical record are both inspired by God and, therefore, compatible. For Einstein, light and the movement of light was the variable that brought scientific understanding to creation. Light, that which is without mass, is pure energy, and pure energy produces mass (matter) and the distribution of mass and the creation of time and space.

Einstein recognized that light and the velocity of light act independent of the motion of its source, thereby confirming his theory of relativity. It is the Bible that confirms the source of light.

Einstein's theory of general relativity led scientists to predict and later discover the radiation afterglow and galaxy seeds that verify his assertion that time, space and substance are interdependent (relativistic physics) and have a beginning.

Einstein's theory and resulting discoveries are proof to many scientists that, not only did the universe have a beginning and is expanding, but also that the universe began as an act of creation.

To hold to the illusion that random natural forces created the world now contradicts established science. Natural laws of nature were created at the same time the world was created. Time, space and substance are interdependent, and nothing precedes them or is able to precede them—except the creator of time, space and substance.

In 1929, Albert Einstein was quoted as saying:

> "We are in the position of a little child entering a huge library filled with books in many different languages. The child knows someone must have written those books. It does not know how. It does not understand the language in which they are written. The child dimly suspects a mysterious order in the arrangement of the books but doesn't know what it is. That, it seems to me, is the attitude of even the most intelligent being towards God. We see a universe marvelously arranged and obeying certain laws, but only dimly understand those laws. Our limited minds cannot grasp the mysterious force that moves the constellations."

Next . . . the chapter you've been waiting for. Did your ancestors swing from trees? Was man happier before he evolved? Is your cousin a chimpanzee?

Chapter 17

Ape to Man:
The Still-Missing Link

The obvious temptation was to title this chapter "Man to Ape." In fact, today some are ready and able to make the case that man was of greater intellect in past generations.

While ivory tower professors are often inclined to flaunt their superior knowledge before wide-eyed students, one would expect that they above all would be the most inquisitive when it comes to a theory with so many flaws. Fortunately, all science and research professors are not so "institutionalized."

C.S. Lewis said, "Evolution is one of those theories where imagination ran ahead of science, and the theory of ape to man is certainly the most imaginative." From a Judeo-Christian worldview, though, I am aware that not all share that viewpoint; however, from a position of Judeo-Christian belief, can one condone an ape-to-man theory? And yet, if one is going to be consistent with the theory of evolution, then man must also have evolved from a former creature.

Based on the multiple inconsistencies in the theory as well as downright fraud, one wonders why men of intellect would continue to discuss it, much less accept it, as a viable theory.

Some zoologists accept it not because the evidence is so compelling, but rather, because no acceptable alternative theory exists apart from design creation. The handful of suggested evidence has all been exposed as fraudulent, meager or riddled with alternative and more logical explanations.

Silence from the field of academia is deafening. Their flagrant disregard for logical and/or scientifically observed and verifiable evidence is incontrovertible proof of an agenda beyond logic and science; therefore, you must depend upon your own understanding and not be caught up in imagination.

Without a designer-God, man is free to be his own god. According to evolutionary humanism, all knowledge originates with man and is taken from a universal body of knowledge. Man's goal is human perfection and is achievable through, among other ways, mantras, mystic breathing exercises and silence. I kind of like that last one.

Unfortunately, our mystic friends, while hoping to gain progressive change, have overlooked the fact that evolution is a theory of death and bound by the natural laws of entropy and decay.

In addition, according to C.S. Lewis, "Evolution cannot and does not explain the origin of life, the variations of life or the origin and validity of reason." Evolution has graduated from simply a theory of descending life-forms into a theory that seeks to improve all forms of existence as witnessed by today's climate fanatics.

We are now to believe that reason, virtue, poetry and all forms of civil society evolved to man from an ape (or another monkey creature). In other words order evolved from chaos. C.S. Lewis states that we are expected to believe that reason is simply the "unforeseen and non-intended by-product of a mindless process at one stage of its endless and aimless becoming." Is it logical to believe that "small, chaotic or feeble things perpetually turn themselves into large, strong and ordered things? Is vice only underdeveloped virtue and love an elaboration of lust?" Lewis questions.

Man has determined to let imagination fill in the blanks when logic and rational thought give inconvenient answers. Thinking is not, nor could it ever be, the result of materialistic changes. Thinking is not matter. Human reasoning is not explained by a mystic body of knowledge, conditioned responses or animal instinct.

Humanism is not new. It began in the Garden of Eden when Satan told Eve she could be like God. Not even Darwin suggested that man came from apes. It was promoted by an anti-god movement and blindly accepted by society. Charles Darwin wrote, "Would anyone trust in the conviction of a monkey's mind, if there are such convictions in such a mind?"

Defining the Debate

In 1925 the Scopes Monkey Trial led to the teaching of evolution in the public schools; gradually six-day creation was marginalized and the theory of evolution began to replace it. For a time teachers and preachers alike attempted to explain creation as a part-time and part-God intervention; however, by the late 1950s all textbooks and public classrooms taught evolution as a proven theory of creation. In 1963, the Supreme Court banned the Bible as a teaching tool in public schools, and soon God and public prayer were officially expelled from the classroom.

On September 11, 2001, close to 3,000 people were killed by terrorists. On that same day over 3,500 babies were killed by abortion. Few see a problem with that as they have learned from school textbooks that a fetus is more like a fish than a human at certain stages; thanks to fake drawings (still in textbooks though admitted fraudulent) used to show the similarities between species of pigs, dogs, fish and humans.

"Yes, I lied, but everyone does it," said Earnest Haeckel, German professor of embryology, when in 1875 he was convicted of fake drawings of embryos that attempted to show similarities between vertebrates. In addition those supposed gills he pointed to in human embryos turned out to be developing bones in the human ear. These same drawings are still being used by evolutionists to prove their embryological theories.

Earnest Haeckel said, "spontaneous generation must be true not because it has been proven in a laboratory, but because, otherwise, it would be necessary to believe in a creator."

Why should creatures who believe themselves evolved from apes not behave as creatures evolved from apes? Public school discipline and parental discipline are a thing of the past. Children now set the family's agenda as if they had never heard the Commandment "Honor thy father and mother" (Ex. 20:12) or "Children obey your parents in everything, for this pleases the Lord" (Col. 3:20).

"This is what the Lord says—He who made you and formed you in the womb . . ." (Isa. 44:2)

Scientists will soon be given governmental permission to use the tissue (stem cells) from aborted babies in research intended to "save lives." More teachers will be fired for mentioning design creation in their classroom while Christ is removed from Christmas and replaced with a winter holiday. Evolution is fast on its way to becoming the state-approved religion.

Who Has "Monkeyed" with the Evidence?

It is with some effort we shall attempt to explain the real truth as we journey through the mythical evolution of monkeys to man (Homo sapiens). Even the editors of the Random House College Dictionary have fallen in line with the monkey-to-man theory as they define Homo sapiens ("sapiens" meaning wise) as the single surviving species of the genus "homo" and of the primate family to which it belongs.

Immediately one should ask the question why Homo sapiens are the only evolved species that can reason. Of course, Darwin and Hitler thought that Jews and blacks were only partially evolved. Over millions of years and the theoretical hundreds of primate-type creatures, why only one Homo sapiens? Except for skin color and differences based on climate and diet, all humans are virtually the same.

Even that view has not always been held by the evolutionist community. As late as the early 1900s the common view among evolutionists was that the oriental race evolved from an orangutan, blacks evolved from gorillas, whites evolved from chimpanzees and Jews evolved from gibbons. Now, doesn't that elevate your faith in evolutional science?

Thomas Huxley wrote that "No rational man can believe that the average Negro is the equal of a white man." Japanese at one time believed themselves superior because they had less body hair and were, therefore, farther away from the ape. Teddy Roosevelt wished he could stop the Indians from breeding. Mussolini thought that Italians were superior, and Germans thought Germans were superior.

Hitler created his own species chart:

1. Nordic . . . . pure Aryan
2. Germanic. . . . . predominantly Aryan
3. Mediterranean . . . . slightly Aryan
4. Slavic . . . . half Aryan, half ape
5. Oriental . . . . slightly ape preponderance
6. Black . . . . .predominantly ape
7. Jews . . . . close to pure ape

Do you ever wonder how the German people were able to participate in the brutal extermination of Jewish men, woman and children less than 70 years ago?

The Search for a Missing Link—Fame and Recognition to the Winner

At the beginning you should be warned that the entire evidence one might expect that proves or even suggests the evolution of man from a line of apes is thin. Evolutionists have laid out a five- to six-million-year narrative outlining their rendering of the ape-to-man theory based on assorted skeletal bones, cave drawings and primitive tools. They have used the same questionable dating formulas and in many cases falsified evidence and concealed conflicting evidence.

Many evolutionists insist that the most natural ancestor of man is the chimpanzee based on the fact that the chimp has 48 chromosomes and man has 46 chromosomes. The only problem with that assumption is that the tobacco plant also has 48 chromosomes, which may explain why some humans become addicted to tobacco, but does not answer the question of ancestral descent.

Evolutionists are in general agreement that the Stone Age (6 million B.C. to 10,000 B.C.) is the period in which humans evolved from monkey or an intermediate monkey-creature to man (what is presently referred to as the Neolithic period through the later Bronze Age and Iron Age). This agreement comes, not based on evidence or discoveries, but rather on the evolutionary belief that it would take six million years for a human to evolve from a monkey or pre-monkey primate.

Evolutionary paleontologists and archeologists have sought to discover those missing links that would provide an evolutionary chain of descent from monkeys to humans in order to prove their assumption.

Why monkey to man? At first only because they looked alike, until later initial DNA studies indicated a close genetic relationship between apes and man. I have no disagreement with the notion that some men look like monkeys and some monkeys look like some men; so instead of disagreeing on that point let us discuss the DNA evidence.

For years evolutional scientists have said that apes and men are only a couple of gene sequences apart in their DNA. What has been more recently discovered is that geneticists were only mapping the axons used for coding protein molecules rather than intones, which make up the coding for major cell assembly, function, maintenance and repair. What was first considered as "junk genes" by Francis Crick is now being considered as the most crucial building blocks of life. While humans and chimpanzees share a common 20-amino-acid strand, DNA and RNA nucleotide structure displays a 48-million sequencing gap. According to Francis Collins, this approximates a 96 percent DNA similarity.

To change the structural unit of only three DNA or RNA nucleotide strands would destroy the life-form. Science, which is so committed to the evolutionary theory, has invented a more recent explanation of how nucleotide might have randomly lined up in past history through a process of "gene duplication"; however, no such mechanism for gene duplication has been discovered. Gene duplication is assumed because a common ancestor for man and apes and all other species is assumed.

While this is only one more example of agenda determining conclusions, it must be pointed out that even if only small DNA differences exist between chimpanzees and man, the same could be said for the DNA differences between a dog and man. Humans, chimpanzees, mice and dogs all share a similar gene sequencing that codes protein, though random DNA segments vary between genes as much as 60 percent. Would it not be logical to conclude that all life-forms are designed to exist in a common ecological system and that their DNA must be designed accordingly for structure, survival and propagation?

The Search Continues

Thousands of fossilized monkey and human bones have been discovered and catalogued. Radiometric and radio carbon dating have been performed in some cases. Artifacts discovered in the same locations have undergone obsidian hydration dating with which volcanic glass (obsidian), used by early man for arrowheads, knives and small tools, is dated according to the diffusion of water (hydration) that occurs in the artifact over time. And naturally the geologic location, discussed earlier, is used to date many bones and artifacts.

Each dating formula has its own inherent problems.

1. Radio carbon dating is restricted to the presence of C14 in formerly living samples and is based on an assumption that the atmosphere today is the same as that atmosphere projected thousands of years ago. Most evolutionary scientists are convinced that the earth experienced an ice age in the not too distant past (10,000 to 30,000 years ago), and that alone would drastically affect the C14 content in the atmosphere, as well as invalidate any age assumptions.

   Creationists believe that a catastrophic flood 4,500 years ago possibly initiating an ice age or in conjunction with an ice age would have drastically affected the C12/C14 ratio.

2. Radiometric (atomic dating of radioactive elements found in mostly volcanic rock) has produced wildly different dates from the same sample and is subject to comparing present chemical content with content assumptions into the far distant past. A catastrophic flood or ice age would also affect present sample isotopic conditions rendering them invalid.

3. The Geologic Column is based on circular reasoning as columns are dated according to the fossils that are found in each stratum, and a fossil is dated according to the stratum in which it is discovered. Evolutionists believe the columns were formed over millions of years, while creationists believe they were formed rapidly in a catastrophic flood.

4. Hydration dating is limited to very recent years when man used such implements. Even then obsidian dating is greatly affected by age variations, exposure to air, temperature and overexposure to water. Any attempt to apply hydration dating to implements or sites predetermined to be "prehistoric" without an additional dating source is admittedly unreliable.

Allow this writer one final thought before we move to the ape-to-man evidence. One author suggested that some of the traction gained by the monkey-to-man theory was re-enforced by the aggressive and sometimes brutal actions demonstrated by humans, thereby, indicating that men have hereditary aggressive tendencies present in their genes from a past evolutionary animal state.

Many scientific studies would take exception to the notion that monkeys as well as other animals are any more aggressive than man. Common tendencies are observed in all species, including protection of their young, their territory and their mate. Friendly tendencies are also observed as found in tribal and family relationships with both animals and humans.

Relationships among animals are more often typified by practical co-existence as opposed to aggressive combativeness. One might more easily conclude that it is man who displays the more aggressive nature; hence, man would then be the "throwback" and not the prized culmination of evolved primates.

Can an Ape Become Human?

There is no easy way to outline all of the fossilized and non-fossilized evidence; therefore, we shall mention and attempt to explain some of those specific discoveries that have been presented to the public as possible missing links. Bear in mind, thousands of bones, teeth and artifacts have been discovered. Most of these have been categorized according to the previously mentioned dating systems. Where one discovery has been proclaimed a missing link, discovered most often by a well-known archeologist, for obvious reasons, those are the ones on which we will concentrate.

Neanderthals

The first skeleton was discovered in 1856 in the Neander Valley of Germany. It was, as were later discoveries (more than 300), described as humans with curved backs. The assumption was made that these were human ancestors beginning to stand upright.

The average heights of later skeletal findings were exactly as one would expect human skeletons to be: 5'3" to 5'9" for males and 5'1" to 5'3" for females. A skull was given to nine artists and nine different renderings were received. Later a prominent dentist examined the skeletal remains and determined that these were old people with diseases—arthritis and vitamin D deficiencies. Evolutionary dating had yielded ages of 30,000 to 600,000 years old.

Similar skulls and skeletal remains have been discovered all over Europe, Iraq and Russia. A protruding nose and forehead is the most obvious facial characteristic used to label the findings.

A recent History Channel documentary has now stated that the evidence demonstrates that Neanderthals are Homo sapiens with the same brain size and burial techniques as humans today, and are probably not more than 6,000 years old. Others have suggested that some are humans who lived 200 to 300 years ago. They noted that some people living today including certain Eskimo tribes and Mongolians have similar physical features.

The Java Man

Accounts vary as to the events and timeline of what was introduced to the public as a major missing link, Homo erectus. In 1892, Dr. Eugene Dubois, a Dutch physician discovered a thigh bone on the island of Java, Indonesia, that appeared human. A year earlier, Dubois had discovered a large skull cap and three teeth in the same general area. There was no reason to believe the thigh bone, the three teeth and the skull cap came from the same creature; however, Dubois made it so.

One later authority identified two of the teeth as being from an orangutan and the other from a human; nevertheless, textbooks and magazines proclaimed Java man to be the long-awaited "missing link."

For 30 years Dubois failed to mention that he had discovered two human skulls in the same geologic area.

Later articles said that Dubois had renounced Java man as the skull of a gibbon and the thigh bone of a human; however, the damage was already done. The Java man still appears in textbooks as an example of Homo erectus.

Many evolutionists believe that the sheer magnitude of fossilized discoveries, though none clearly distinguish a verifiable missing link specimen, speaks for itself. With so many discoveries of various pieces of the puzzle, the puzzle becomes reality. What has happened in reality is that drawings were made according to preconceived imaginary links that must exist if man evolved from an ape. Thousands of fossilized bones, therefore, have been dated and labeled according to these imaginary drawings based on preconceived imaginary links.

The Piltdown Man

In 1912 the New York Times wrote "Darwin's Theory Proved True." According to many, Charles Dawson (not Darwin) began his research for the missing link in 1908, and by 1912 claimed to have found the missing link in a gravel pit at Piltdown, a village near East Sussex, England. What ensued was a 40-year hoax not exposed until 1953 as a forgery. Piltdown was listed, along with Neanderthals, as an example of every hominid life.

Workmen had discovered a fossilized skull fragment a year earlier and when Dawson arrived he found other fragments, according to him, in the general vicinity. Dawson later discovered the jaw bone of an ape, which he filed down so the two head fragments would fit together. The jaw bone was later determined to be from a 10-year-old orangutan with teeth filed down and age stains added to make them appear ancient.

The hoax was actually exposed in 1923 by an anatomist, but it took another 30 years to remove it from textbooks. Later, the filed-down teeth were determined to be from a chimpanzee. The Piltdown man was actually a composite of two apes and the 600-year-old remains of a man.

According to one determined evolutionist, "this disqualification of the Piltdown skull changes little in the broad evolutionary pattern." Perhaps not, but this deceptive pattern has continued to broaden.

Dawson was also responsible for at least 38 other forgeries including what was presented as teeth from a reptile/mammal hybrid that became known as the "Piltdown turkey." It was later smuggled out of China and sold to a private collector in the U.S. In 1999 National Geographic magazine exposed it as two fossils stuck together with glue—the fossilized remains of a large feathered bird and the tail of a dinosaur.

The Nebraska Man

What was originally heralded as a possible link to a higher primate of North America (found in Nebraska), actually turned out to be the discovery of a fossilized pig's tooth. From that one pig's tooth, an artist's rendering appeared in the Illustrated London News in 1922 in a two-page spread depicting an ape-like man and his wife hunting for food. The artist explained that he modeled his drawing after the well-known Java man.

Nebraska man, according to one writer, later became known as pig man and misidentification continues among paleontologists, though it does not discourage their efforts.

The Tatung Baby

The Tatung baby was discovered in 1924 by Raymond Dart in a South African limestone quarry. This single-skull fragment not only tells us that he or she lived in caves with others in the clan but was agile enough to fashion primitive tools and hunt small animals (fairly talkative for a skull fragment) just because its brain cavity was the size of a small gorilla and resembled that found by the Leakeys' Nutcracker man.

The Nutcracker Man

A single skull discovered by anthropologists Louis and Mary Leakey in 1959 seemed to indicate that this assumed forerunner to humans ate mainly nuts to survive during its transitional lifespan. Others as late as 2008 have challenged that theory based on a re-examination of the teeth and jawbone.

National Geographic, sponsors of this expedition, used creative imagination when it filled in the missing body parts by allowing this single skull to walk upright and covered it with extensive body hair necessary for a skull-person to look transitional. To some who viewed these drawings, this skull person's eyes seemed almost human in its pensive appearance as he stood with other members of his imaginary family.

More recent studies have assigned this skull to a now-extinct group of apes in the southeastern regions of Africa.

Lucy

Lucy was the discovery of Dr. Donald Johansson and his team in 1975, two weeks before his archeological research grant was to expire.

Johansson began exploring the Hagar region of Ethiopia in the early 1970s and found it rich in fossilized remains. Later he and a colleague began exploring the Afar locality and by piecing together numerous bone fragments from that region came up with a 40-percent-complete skeleton they named Lucy, supposedly after a Beatles' song. Later, in a different area of Afar, one mile away, they discovered a hominid arm bone and several pieces of jaw bone.

The team decided that all of the bone fragments were from one individual as, according to them, they did not find any two pieces alike. Lucy was dated by radiometric measurements of potassium displacement to argon gas performed on volcanic ash in the area where some of the bones were discovered. Based on the geologic location in which parts of Lucy were found as a starting point, she was dated 3.2 to 3.8 million years old.

I must remind you that radiometric dating requires a second source as a template to begin a measurement. If the geologic stratum is the second source and its assumed age is already predetermined, that becomes the starting point in which to structure a radiometric measurement.

The contradictions began soon after Lucy was announced, though many evolutionists still remain faithful to the integrity of this pieced-together skeleton.

Lucy measured three feet, eight inches tall and, though the skull was crushed, she appeared to be a collection of fragments from a common chimpanzee. Textbooks show a half-human and half-ape -like skull (the ape portion was found a year later).

The knee joint is what evolutionists insisted allowed Lucy to walk upright. Actually the knee joint was found one and a half miles away, though, when presenting his findings, Johansson said it was nearby.

The St. Louis Zoo put human feet with a big toe separation on Lucy even though no foot bones were found. It was never mentioned that normal human footprints were also found in the ash next to some of the pieces of Lucy. The zoo also added dark skin and hair to its mockup, which was pictured in National Geographic in 1979.

Many evolutionists dated Lucy in the Cro-Magnon era. Others have suggested that Lucy might be a normal human based on the human footprints found in the ash next to parts of Lucy. There has been nothing, however, in the fossil record to indicate that Cro-Magnons were anything other than humans.

The Turkana Boy

Found in Kenya in 1984 and originally dated at 1.7 million years ago, this skeleton stood five foot, three inches tall and was presumed to be the remains of a boy nine to 12 years old. He walked fully upright and had a brain cavity similar to humans today. His overall body stature, weight and proportions were similar to modern-day Maasai living in Kenya. He was undoubtedly a normal human boy incorrectly dated.

Today many archeologists believe that Homo erectus, Cro-Magnon man, Neanderthals and Archaic Sapiens are all modern human types. Erectus and Homo sapiens are the same, though thousands of fragment pieces have been classified as transitional ape-to-man evidence. Some originally radiometric-dated several million years old have more recently been radio-carbon dated at 12,000 to 35,000 years old and, when they are calculated with allowances for a catastrophic flood, they can be recalibrated at less than 4,000 years old.

Dating is highly inconsistent because starting points and sample integrity is unreliable. Skeletal deformities can be attributed to poor diet as well as people living longer lives prior to the flood and ice age. Because so few morphological differences exist when the entire fossil record is considered, there is no compelling reason to believe that all fossilized specimens are not either completely human or completely ape.

Even if radiometric dating was reliable and accurately dated samples up to six million years old, according to some geneticists, this still does not allow for enough evolutionary-long periods of time necessary for the slow incremental changes (gradualism) that would be required to evolve an ape or common ancestor into a man.

Cave drawings and stone tools are not evidence of ancient cavemen as some people today still live in caves and work with stone tools. The early American Indian, people of New Guinea, Borneo, Central America and parts of Africa has cave dwellers who continue to use stone implements. For a short time, King David of the Old Testament lived in a cave.

Museums, National Geographic and outdated textbooks still insist on telling us that man evolved from an ape creature and at one time lived in caves based on scattered or fraudulent evidence. Most states have laws requiring textbooks to be accurate. These laws are overlooked when it comes to the religion of evolution.

Professor Protsch's Many Forgeries

While seeking fame and recognition, Professor Reiner Protsch von Zieten, once a distinguished German archaeologist, discovered what he said to be a missing link between Neanderthals and humans. It was the professor himself who falsified his own carbon-dating test as well as, what was later learned, tests on many other supposed "stone age" relics over a 30-year period.

The archaeological community had once again jumped to rewrite their ape to human paradigm based on a single and later-to-be-discovered fraudulent bit of evidence. Protsch was also found to have falsified the dating on Hahnhofersand man, a supposed Neanderthal carbon-dated around 30,000 B.C., now dated 7,500 years old; the Binschof-Speyer woman, who lived 3,300 years ago and not 23,300 and the Paderborn-Sabdenab dated at 27,400 B.C. who had actually died in 1750. After decades of fraudulent and mistaken dates, Professor Protsch was suspended from the University of Frankfurt in 2005.

What won't a professor do for attention? Sadly in a field of study where so many conclusions are made based on indefinable variables, and where instant notoriety awaits the one who gives evolutionary support to Darwin's origin of the species, fraud has become common place.

The Devil's Archeologist

Earning his nickname, "God's hands", Japanese amateur archeologist Shinichi Fujimura was caught digging holes and burying objects that he later dug up and heralded as major discoveries. Based on his buried findings, textbooks were rewritten and the Japanese Paleolithic period extended 300,000 years. After a news reporter published pictures of his fraud in November 2000, Fujimura said the devil made him do it.

The Dikika Baby

What appears to be the more recent fossilized discovery claimed to be an ancient forerunner of man was unearthed in 1999 by a team of fossil hunters in Ethiopia. National Geographic was quick to label this as the most important discovery since Lucy. It was dated at 3.2 to 3.5 million years old (seemingly corresponding with Lucy).

In fact, this was a far more complete skeleton than Lucy. The Dikika baby was found somewhere around 2.4 and 6 miles (accounts differ) from Lucy (claimed to be an adult female) and was the remains of a three-year old infant. Prior to this discovery all the fossilized remains of a baby, according to National Geographic, could have fit in a diaper. She was named Dikika based on the location in the Dikika Valley of Ethiopia in which she was found.

Why the excitement? What paleontologists claim is that the Dikika infant, while displaying the normal bone structure of a three-year old chimpanzee, including the same brain cavity, also possessed a partial knee and a nearly complete leg and foot that suggested the ability of walking on two legs (bipedalism). The Dikika baby still possessed the separated big toe and gripping fingers of an ape.

The rest of this article that appeared in National Geographic in November 2006 is all imagination based on "what must have been" if man truly evolved over six million years ago from an ape creature.

1. **Why Was the Dikika Baby Dated at 3.2 to 3.6 Million Years?**

    The answer to that question is purely speculation. There is no indication that radiometric dating was even attempted, and even if it was, the sandstone surrounding the skeleton would be completely unreliable. When the skeleton was discovered the face was showing, which indicates loose sandstone in the surrounding area—archeologists admit this is proof of flooding.

    The most logical answer is that the Dikika infant was dated based on Lucy's date and according to the geologic location in which it was discovered. If ape-to-man evolution is assumed, then dates are affixed according to a preconceived evolutionary order.

2. **Why Was It Assumed to Be a Forerunner of Humans?**

    There is absolutely nothing on which to base such an assumption except the humanlike knee joint. In all other respects the Dikika infant is an ape.

3. **How Trustworthy Are the Conclusions Reached by National Geographic and These Ethiopian Paleontologists?**

    a. In the same location were also found fossilized bones from elephants, rhinoceroses, antelopes

and hippopotamuses, none of which had evolved beyond what they have always been.
b. Even if radiometric dating were reliable, the exposed sedimentary layer in which Dikika was found would have compromised its chemical integrity, thus not allowing for radioactive, mother-daughter ratio testing.
c. The geologic column cannot be used as a reliable source for determining the starting point of radiometric dating as fossil content is jumbled and not uniform.

Is It a Fraud?

No reason exists to believe the skeleton is not real; however, every reason exists to believe that evolutionists have, once again, created a mythical ape-to-man chronological timetable to which all fossilized human and monkey bones must fit. Often it is fossilization alone that prompts evolutionists to believe that a sample must be very old.

Why would it be assumed that a monkey who is climbing trees and generally living a happy, carefree monkey's life would consider (if a monkey could consider) evolving into a human to be a step forward in evolutionary descent? Is that something a monkey can imagine with its 60-percent human brain capacity and two additional chromosomes?

Why is it that only the monkey-man, who supposedly evolved to walking upright, is the only primate who can reason? Does walking upright have anything to do with conscious awareness? With a suggested millions of years of fossil history, according to evolution, why are there not numerous clear lines of transitional human life-forms? If the Geologic Column is a true indication of 600 million years of evolved life, where are the evolved transitional life-forms for all species including man?

What About the Bonobo Monkey?

Often called the pigmy chimp, this monkey still exists along the Congo River and is, perhaps, the most human-acting of all apes. It walks upright and is probably the best explanation for the Ethiopian Dikika skeleton. All other ape-like characteristics appear to be the same.

Whether this is true or not, what the fossil record does show is that all fossilized life-forms ever discovered are either extinct or still exist in their present form. Some evolutionists point to the Bonobo monkey as the common ancestor of both Homo sapiens and chimpanzees, which makes absolutely no sense at all. (1) All three exist today in their present form; (2) All three have remained in their present form.

The Search Continues

Evolutionists will say . . . "due to the special circumstances required for the preservation of remains, only a very small percentage of transitional life-forms can be expected to be represented by discoveries."

That of course, is completely contradictory to the whole theory of slow changes over long periods of time. Logically, a wealth of transitional discoveries should exist, especially with so many anxious evolutionists searching for them.

Evolutionists will say . . . "Progress in research and new discoveries will continue to fill in the gaps." The gaps, however, are in the theory of evolution as well as the existing record. What one might wonder is why not wait for those discoveries and progress in research and let that determine one's theory, rather than letting one's theory be used to force conclusions about discoveries and research that do not exist?

Nothing, I repeat, absolutely nothing has been discovered that demonstrates even a slight evolution from an ape creature to humans. It may make for great movies, but it is not real.

If evolution is true, then man must have evolved through the same slow process as all other life-forms from a common ancestor. Imagine the weird transitional life-forms that would create. Fossilized transitional life-forms should far outnumber those life-forms in which the "conditions of life" have been "perfected" according to Darwin.

Now You Must Reason

In this lesson we have concentrated on monkey to man. Why is man the only life-form with conscious and moral awareness? Why is man the only Homo sapiens who reasons?

If evolution over long periods of time is not intellectually, scientifically, observationally, philosophically or mathematically possible, what's the alternative explanation? And the answer is . . .

No one has to accept the real truth, the supernatural answer, the Word of God. That is a choice each person must make. But, be perfectly clear, one's choice to hold on to a godless (thank you, Ann Coulter) theory of creation of which death is the dominant theme, of which one life-form must die in order for another to be perfected, instead of accepting a theory of designed life, is a sad, sad commentary on the condition of one's heart as well as one's ability to reason once alternative evidence is made known.

This is what Paul said to the church at Colossae, "See to it that no one takes you captive through hollow and deceptive philosophies, which depends on human tradition and the basic principles of this world rather than on Christ" (Col. 2:8).

A Personal Opinion

As we approach the end of this study, allow me to relate what we have learned based on what it reveals about the theory of evolution to what is happening in the world today.

Assuming you are paying attention and are old enough to look back a few years and assess the rapid change taking place in the world and in America today, you must realize that choices are being clearly defined in both the political and religious arenas. Not even "Future Shock" could have predicted what is happening before one's very eyes.

Politically, the choice between personal freedoms as opposed to governmental control is the issue. The writers of the American Constitution warned us of this possibility; however, I suspect few in America ever imagined they would be forced to make that choice, assuming that choice still remains.

More importantly, the choice of one's acceptance or one's rejection of the biblical God of creation is also being clearly revealed. I reject the notion that rejection of God is an acceptable scientific and logical decision.

It is a choice based on clearly revealed alternatives. It is almost as if God is bringing final clarity to mankind's spiritual choice by filtering out all the conflicting excuses. It is as if God is saying, "Accept or reject, but don't excuse your rejection of me based on anything other than your unwillingness to surrender to me."

The Revelation of Jesus Christ to the disciple John is upon us, and most of the world and the church are oblivious to the impending tribulation.

Are these the final days? Many believe they are. Either way, they are your final days and your opportunity for acceptance and surrender.

Chapter 18

Evolution Fails to Answer

Perhaps Christians have been waging battle in the wrong arena. In our enthusiasm to stand upon the Bible, biblical faith and biblical principles, we have often directed our discontent towards secular education and secular institutions rather than focusing on the church, the body of Christ. Knowing that faith is a choice influenced by evidence and logic but not determined by evidence and logic, born-again believers should be inclined to follow Christ's command of growing up the Church (making disciples).

The truth is that we have often failed in the church arena. The Bible tells us to "train up a child" and we have more often spent our time carrying protest signs in a secular world and allowed the world to "train up a child."

Those Pesky Questions

1. Why is it that so many legitimate scientists, trained in the same schools and working side by side with other scientists, do not accept evolution in fact or theory? Is it because once they step outside the laboratory and into a church pew they become brain dead to reality? Perhaps it is because a miracle has taken place in their lives and they, with "eyes to see", recognize the hand of a creator.

2. How do evolutionists explain why creation is the only random act of nature originating out of chaos while everything else in the universe clearly displays organized design? Even the building of animal tissue is in harmony with the growth of surrounding tissue. A network of arteries and capillaries deliver oxygen to cell tissue for metabolism while the veins take away de-oxygenated blood and waste byproducts. DNA molecules are repaired and corrected instantaneously by an enormously complex system of enzymes, some of which must be acquired in nature, outside the body. DNA information could have never originated itself. Though its design has now been decoded by man, it is that complex design that rejects any possibility of chaotic chance assembly.

The human eye automatically focuses over 100,000 times a day. Two eyes allow for coordinated focus to one point while eliminating double vision. The eye is self-healing and self-cleaning. Science can observe optical design and medically treat some optical malfunctions; however science cannot create a single optical cell.

3. Why is the theory of evolution accepted, if the statistical probability of it occurring is zero? How many other theories could have survived into the 21st century, when every biological and mathematical probability says no way? It is not that there is only a slight chance of occurrence, according to mathematical probability—there is no chance of its occurrence, no matter how long the waiting period.

Could the universe form itself? See if you can follow the calculations of one mathematician: Assume that the universe is five billion light years (5×10^9 powers) in radius. A light year is the distance light travels in a year while moving at a speed of 186,000 miles per second. I note that while this figure is well within the range of estimates as to the size of the universe, most scientists now believe the universe is still expanding.

If the empty spaces in the known universe were filled with the smallest known particles (electrons), those particles would number 10^{80} power and fill the universe.

The probability of life appearing as a result of an energy flux event created by required electrons is $10^{130} \times 10^{20} \times 10^{20}$ power or a total of 10^{450} powers—well beyond the number of electrons available in the universe and required to supply the potential energy needed for chance creation of the universe.

Based on known cosmological certainties, evolutionists have been forced to admit that the universe had a beginning. Not only did this fact put an end to the then-existing evolutional theory that matter was eternal (at one time held by Albert Einstein), it also became the nail in the coffin for the theory that "anything is possible if time or the passage of time is not a factor." If, indeed, the universe, including the earth, had a beginning, time is a factor and the evolutional timescale itself, as a limiting factor, destroys any possibility of slow changes over unlimited time. The creation of time and matter has confirmed the zero possibility of life from non-life and evolutional descent.

4. Who or what supplied the information necessary for life to begin from non-life and regenerate itself over and over again? There is no known mechanism whereby life might randomly occur, even though thousands of scientists have sought to discover one.

Einstein recognized that information, including DNA, has nothing to do with matter or energy. Archeologists make decisions daily as to whether an object was produced by natural causes or by humans. When they find a silver idol the thought never crosses their mind that this idol was produced over long periods of time by chance environmental circumstances and rogue bursts of energy; and yet, many are convinced that the universe and all life appeared by happenstance. How can society extol the creative successes of mankind while attributing the monumental complexities of creation to chance?

Man can produce synthetic soil but no scientist has ever built a synthetic seed. Every seed is predesigned with all the necessary information for germination and growth after its own kind. Every plant is endowed with the genetic formula to reproduce itself. An equally amazing miracle, beyond predesigned genetic coding, is the environmental interdependency required by every seed from the beginning of new life. Seeds require water, soil, light, the proper atmosphere and pollinators in order to survive. One author reminds us that even flowering plants, not just food plants, provide needed oxygen and, at times, natural herbs and medicines.

The fact that God created all things good does not simply mean that God created them pleasing to the eye, which they are, but rather, that He created all things based on interdependence and, sometimes, symbiotic relationships (mutually beneficial association of two dissimilar life-forms). Careful design, ecological codependence and unique symmetry are good to God.

Snow crystals contain millions of water molecules. Every crystal exhibits its own unique symmetrical pattern. The energy available in one tree is enough to power the United States for 10 years, if science were able to convert its mass to useful energy. Solar bodies display such amazing differences that it is difficult not to believe in a God of unlimited diversity, complexity and beauty. That is what God meant when He said it was good.

5. Why is evolution not observed today or in the fossil record? What the fossil record shows is that humans as well as all other life-forms appear abruptly in their present form and either still exist today or have become extinct. The observed evidence indicates that evolution stopped before it even began.

In no place is the lack of transitional life-forms more obvious than in the fossilized and non-fossilized centerpiece of animal evolution—the dinosaur. According to the evolutional time table, dinosaurs appeared approximately 150 million years ago. With as many fossilized dinosaur bones having been discovered around the world, evolutionists are still unable to explain why transitional dinosaurs don't exist. The British Museum has the largest collection of fossils in the world. It does not include any transitional life-forms. Every fossilized or non-fossilized dinosaur bone or skeleton has revealed all to be the same in kind and form. All appear at once without any sign of descent and, according to the evolutionary calendar, continue in their same form for the next 90 million years.

The geologic record is jumbled with dinosaurs at virtually every theoretical stratum. Dinosaur footprints have been discovered alongside human hand and footprints, considered impossible by evolutionists.

There is no slow and gradual development of any species, including dinosaurs.

6. Why have so many people accepted a-biogenesis (the production of life from inanimate matter) without any proof of its possibility? Rarely spoken of by evolutionists is the fact that eight to 10 of the amino acids required for human growth must be obtained from food outside the body as the human body is unable to synthesize them on its own. Four are especially required by children as their bodies begin to develop. To construct, synthesize, carry and eliminate chemicals in the body, 55,000 enzymes are needed by the body and must be received from plants and animals. Not only does this eliminate a-biogenesis as a theory, it demonstrates, once again, the necessity for a fully functional organism created in a fully functioning ecosystem.

7. Who or what provided the material for the Big Bang? What caused that material to explode? How could that fluid or gas change into a thickened gravitational mass and produce the stars, galaxies and planets?

Muddy water full of righteous chemicals, long periods of chaotic time, miracle energy sources that promote favorable circumstances, complexity from aimlessness, reason from the irrational and conscious awareness from dysfunctional, mutant genes—have I forgotten anything?

Many evolutionists reject the supernatural out of hand. To those evolutionists God is unimaginable. The problem with that belief is that every sacred principle of evolution is 100-percent scientifically unimaginable.

Evolving Theories

Evolution is a theory supported by evolving theories. There is the theory of a-biogenesis, life from non-life, which has never been replicated. There is the theory of an ancient earth based on Lyell's theory of geologic layers formed by uniform erosion over long periods of time, though the layers are far from uniform. There is the theory of evolved life from a common ancestor, based on the morphological similarity of animals, though it is contradicted by genetic sequencing gaps and denied crossover among kinds.

There is the theory of descending life from smaller, less complex life-forms to larger, more complex life-forms, based solely on limited and scattered fossils discovered in the Geologic Column, as later discoveries have revealed a jumbled fossil record. There is the theory of man evolved from an ape creature, though genetics deny its possibility and no intermediate links have been discovered that were not fraudulent or highly questionable.

There is the theory of a chance mythical exploding dot (or no-dot) creating the universe, though natural law and cosmological discoveries deny its possibility. There is the theory of mutating genes advancing life rather than degrading life. There is the theory of a chemically blessed pool composed of the necessary DNA, RNA, amino acids, enzymes and other non-living molecules necessary for life to form itself into the first living cell, though the entire universe could not contain the required molecular particles necessary to create one living cell by chance.

There is the theory of human consciousness being the result of a genetic mutation based on monkey-man's need to survive. Every theory is supported by more theories, and not one theory is supported by observed and reliable facts or scientific proof—only theories.

Chapter 19

A Final Thought

> Michael Jackson died today, June 25, 2009, following a massive heart attack at the age of 50. To Charles Darwin, Michael would have been considered less than human, but to his fans and the music industry he was a genius.
>
> Farrah Fawcett died two hours earlier, finally succumbing to the cancer that claimed her body. Because of her skin color and Aryan heritage, she would have been considered by Darwin as more evolved than Michael and, therefore, deserving of survival in the struggle for life.

Many evolutionists would be appalled at the racist dogma inherent in the theory of chance creation and survival of the fittest. Humans are not excused from evolving.

If conscious awareness is only one or two gene mutations higher on the evolutionary ladder, why would one consider a human of more value than any other animal? And for that matter, why would one consider any human of more or less value than any other human when it comes to survival?

Mike Huckabee said that in spite of Michael's human flaws, he deserves our respect, but respect and human value are moral attributes that have no place in a happenstance world. Can value and respect be the byproduct of an aimless struggle for survival?

If you have managed to read this entire study, you have at least been exposed to more alternative evidence to Darwin's 19th-century evolutional theory than the vast majority of Americans. Once academia decided that evolution was a proven fact (no matter what the contradictions) this theory took on a life of its own. I can think of no other comparable theory to gain such widespread, and yet, divisive acceptance.

I trust you have also realized that for many evolutionists, belief has set their agenda, and their agenda has often biased the evidence and their biased evidence has resulted in an unscientific theory being proclaimed proven fact.

What I, along with other people of biblical faith, object to is the labeling of the theory of evolution as proven, which it is definitely not by any true scientific criteria.

For the Christian—one who believes in Jesus, the Son of God—no matter how well-meaning one's attempt might be to justify evolution with a supernatural transcendent God of the universe, it is anti-Christ. True science (God-created laws of nature and the universe) do not contradict God's Word; however, God is capable of and has quite often contradicted His own laws of time, space and substance. The incarnation of Christ (God dwelling in human flesh) is the greatest of all examples.

And finally, God never promised that we would be given full understanding of His nature or His Word. In fact, He promised exactly the opposite. Paul longed for the day he would see God "face to face" in order that he might then "know" God even as God knew him (see 1 Cor. 13:12).

I can think of no better scripture to confirm this point than the words Jesus spoke to Thomas following Christ's resurrection. When the disciples told Thomas, one of the twelve, that they had seen Jesus alive after His crucifixion, Thomas said, "Unless I see the nail marks in His hands and put my fingers where the nails were, and put my hand into His side, I will not believe it" (John 20:25).

One week later, Jesus allowed Thomas to do just that, to which Thomas proclaimed, "My Lord and My God" (vs. 28).

Ah, but next come the words of Jesus clearly directed towards you and I today as we struggle with our own doubts. Jesus said to Thomas, "Because you have seen me, you have believed; blessed are those who have not seen and yet have believed" (vs. 29).

Belief in God the Creator and Christ His Son is not devoid of logic and evidence; however, it always has been and always will be based on God's gift of grace and man's free choice of trust and surrender.

Michael died today and, for a time, his memory will live on. Millions will mourn his death, but why? If Michael is only one of 10 or 12 billion Homo sapiens who have accidentally inhabited this planet since random generation of life, his life has no real meaning.

Apart from God's divine creation, life has no meaning. It is as Shakespeare said, "We strut and fret our hour upon the stage and then are heard no more."

James, the brother of Christ, told the early Christian believers to submit themselves to the will of God. "There is only one law giver and judge, the one who is able to save and destroy . . .

you do not even know what will happen tomorrow. What is your life? You are a mist that appears for a little while and then vanishes" (Jas. 4:12-14).

I suppose every writer who has ever attempted to share a commentary on life or religion comes to the end of that project hoping to find that one sentence that will pull it all together and leave the reader with a new understanding or, at least, a satisfying smile.

My conclusion is actually the beginning and the end. It is what gives life purpose, respect and value. It is the fulfillment of every man's hope and the confident anticipation of life beyond this creation. It gives God's meaning to life.

"In the beginning, Elohiym (God) created the heavens and the earth." (Gen. 1:1)

---

*As of this date Michael Jackson's death has been ruled a drug overdose homicide and his personal physician has been charged.

A Children's Story

Hannah's Eyes

Nine-year old Hannah was born blind. She had never seen a sunrise, though she felt its warmth. She had never seen a pigeon, though her father tried to describe them to her every time they visited the park. She did hear their cooing as they scurried for the seed that Hannah always remembered to bring.

One day, following a spring shower, a beautiful rainbow appeared in the sky and her father cried as he feebly tried to explain to his lovely daughter the vibrant colors that Hannah could not see. "God," he said, "made the rainbow and the sunrise and the pigeons in order that we might be able to see Him through His beautiful creation."

"I wish I could see God," Hannah said, as she squinted her eyelids real tight.

Her father tried to explain to Hannah that we don't actually see God but that He is, nevertheless, everywhere around us. "That's okay," she said, "I can pretend."

It was five days before Hannah's 10th birthday when the phone rang. One of the many doctors to whom Hannah's parents had consulted was excited about a new procedure that might bring sight to Hannah's eyes. Though he was greatly encouraged about Hannah's prospects for sight, he warned that the procedure was new, though early tests had proven successful.

As Hannah's parents explained to her what the doctor said, they were careful to caution Hannah about not getting too hopeful just in case the procedure did not go as planned.

"I understand," said Hannah. "God and I talked it over and He said that my eyes were meant to see."

The day that Hannah turned 10 years old, she entered the operating room and two hours later she was out of surgery and in recovery. All the years of prayer and doctor visits now came down to one week while Hannah wore bandages over her eyes as they recovered from surgery.

"If it's okay with you, Daddy, I would like to be in the park when they take off my bandages," Hannah told her father. "That's where you told me about God and the beautiful colors of the rainbow and the bright sunrise, and, of course, my friends the pigeons."

The day came for Hannah's bandages to be removed, and although Hannah must have been excited, she appeared quite calm—far more so than her parents and the doctor who could hardly stop talking, all at the same time.

As the doctor removed the first bandage he told Hannah to keep her eyes closed until both bandages were removed and then he would let her open both eyes at the same time. Now the moment came and the doctor said, "Hannah, open your eyes."

"Oh Daddy," Hannah said, "the colors are beautiful and that must be the sun so bright in the sky."

"Look," said her father, "the pigeons are coming to see you. Can you see them?"

"Yes, yes," she said, "they are waiting for me to feed them. But Daddy," Hannah said, "you were wrong."

"What do you mean, Hannah? I would not tell you anything that was not true."

"Daddy," said Hannah, "I do see God."

With a look of concern on his face, Hannah's father said, "Where do you see God, Hannah? Show me where you see God."

"Everywhere, Daddy, I see God everywhere."

> Jesus said . . .
>
> "But blessed are your eyes because they see
> and your ears because they hear.
> For I tell you the truth, many prophets and righteous
> men longed to see what you see, but did not
> see it, and to hear what you hear, but
> did not hear it." (Matt. 13:16-17)

Glossary

Agnostic	One who doubts the possibility of an ultimate being, God or ultimate knowledge.
Amino Acid	One of 20/23 kinds of building-block subunits of protein.
Amoeba	A microscopic one-celled or simple-cellular animal.
Anomaly	A deviation or inconsistency from the expected rule.
Archaeologist	One who studies historic or pre-historic people or cultures.
Aryan	A descendent of Nordic or Indo-European people.
Astronomer	One who observes the material universe beyond the earth's atmosphere, as opposed to an **astrologer** who seeks to interpret the influence of heavenly bodies on human affairs.
Atheist	One who denies the existence of God or an ultimate, supernatural being(s).
Biogenesis	The production of living organisms as a result of the activity of other living

	organisms as opposed to **a-biogenesis** (life from non-life).
Biology	The science of living matter that deals with growth, reproduction, structures and characteristics of living organisms.
Carbon	The essential chemical element that forms organic compounds in combination with other elements required for all organic life.
Cell	Microscopic plant or animal structure containing nuclear or cytoplasm material enclosed in a membrane (animal) or cell wall (plant). The average human cell has approximately one trillion components.
Christian	A follower of Jesus Christ. One who believes and seeks to follow the biblical teachings of Jesus Christ.
Christology	The study and interpretation of the nature, person, words and deeds of Jesus Christ.
Chromosome	Threadlike bodies found in a cell nucleus that carry genes in linear order.
Cosmologist	One who studies the philosophical and evolutional structure, aspects and origins of the universe.
Creation	The act of creating, producing or causing to exist.
Creationist	One who believes in the divine (God) design and creation of the universe and all life as set forth in the Bible.

Element	Any one of a class of 117 known substances that cannot be separated into a simpler substance as opposed to a compound element or substance containing two or more elements united in a fixed proportion.
Embryologist	One who studies the early stages of fetal development.
Empirical	Derived knowledge from verifiable experiment or experience.
Elohiym	Of the various names given to God in scripture, "**Elohiym**" is actually the plural form of "El" and appears to designate the supreme creator God or Godhead as compared to "**Yahweh**", which appears to designate the God of the covenant. Elohiym implies a God of power as well as involvement in creation out of "**bara**," nothingness; who is by virtue of creation the one true God.
Energy	The capacity to do work, available adequate or abundant power. (Physics) the property of a system that diminishes when the system does work on any other system by an amount equal to the work done.
Entropy	A measure of the amount of energy that becomes unavailable for work as well as energy's tendency towards equilibrium.
Enzymes	Cellular organic substances capable of producing chemical changes in an organism (i.e. digestion), which are produced by the body or ingested from plants and animals.

Equilibrium	A state of balance or rest as a result of equal action of opposing forces or powers.
Evolution	A process of growth, formation or development.
Chemical Evolution	The theory that all chemical elements evolved from nitrogen, which presently makes up about four-fifths of the earth's atmosphere.
Cosmic-evolution	The theory that the universe began as a dot or no-dot with a big bang and will eventually return to the same and repeat the process.
Macro-evolution	The theory that all life began as one single cell or common ancestor and evolved through a process of mutations, environment and unknown phenomena into all the plant and animal species now living or extinct.
Micro-evolution	The observed genetically permitted changes that occur within species of the same kind.
Organic evolution	The theory that all plant and animal life began spontaneously from non-life as a single cell.
Stellar evolution	The theory that deals with the changes a star will go through from birth (which has never been observed) until death (a supernova).

Evolutionist	One who believes in the tenets or doctrines of evolutional creation and descent while denying the supernatural intervention or design of creation.
Evolutional Theist	One who believes God entered the evolutional process in order to achieve creation.
Existentialist	One who believes in man's absolute freedom of choice without any rational or ultimate truth or criteria imposed or serving as a basis for man's choice. Existence alone determines reality and essence of choice.
Extinct	No longer existing, died out.
Fossil	Any remains, impression, track or trace of an animal or a plant that has become mineralized.
Fossilization	To replace organic substance with mineral substance.
Fossil Record	The collection and categorizing of fossils most often related to the evolutional theory of geologic strata.
Fraud	Deliberate deceit or breach of confidence used to gain an unfair advantage.
Gene	A unit of heredity that, along with other genes in a chromosome, controls the growth, development and characteristics of a life-form.
Genesis	From a Greek word having to do with origin or beginning.

Geneticist	One who studies the science of heredity through genetics.
Geologist	One who studies the composition and formation of rocks. In an evolutionary sense, one who theorizes about the physical history of the earth based on geologic or topographical observation.
Gnosticism	Early system of false teaching based on the Greek word for knowledge that proposed the general principle that spiritual understanding can be derived from natural existence as opposed to divine intervention.
Humanism	A mode of thought and subsequent actions whereby human welfare, dignity, interest and needs are elevated to primary importance when making moral decisions or value judgments.
Inerrant	Free from error
Instantaneous creation	Occurring or completed in an instant or a very short space of time, as in a moment
Logic	A method of reasoning that seeks to realize the principles of correct or reliable inferences.
Metaphysics	The branch of philosophy that deals with the nature of existence and reality apart from any concrete or specific reality.
Molecule	The smallest physical unit of an element consisting of one or more like atoms or two

	or more different atoms in a compound element.
Molecular weight	A measurable sum of all atoms in a molecule.
Morphology	Biological term used to describe similar structure or form in plants and animals. Darwin expressed it as "unity of type" when he said that "several parts and organs in different species of a class are homologous".
Mutation	A change or departure from the present type. While evolutionary precepts envision genetic mutations to be advancements in life-forms, in reality, genetic mutations have only been observed as harmful or destructive to life.
Naturalism	(Theology) The theory that all religious truth is derived from natural processes and not from revelation. (Philosophy) The theory that there are no non-natural objects, processes or causes. (American Association of Biology Teachers) "The diversity of life on earth is the result of evolution . . . A natural process of temporary descent with genetic modifications that is affected by natural selection, chance, historical contingencies and changing environment."
Naturalist	One who studies the material world (plants, animals and geographical features), preferably where they have been least affected by man. More recently, natural

history and naturalism have come to mean inferences or conclusions that may be drawn from measured observation and/or natural revelation as opposed to divine revelation.

Nitrogen Gaseous element that constitutes four-fifths of the earth's atmosphere.

Nucleus The core or central part. (Physics) the central core of an atom composed of protons and neutrons.

Ordinary view Darwinian term used to describe the independent creation of each being so as to please a creator God.

Organism A life-form (plant or animal) composed of mutually dependent parts allowing it to maintain the vital functions needed to sustain and reproduce life.

Paleontology The study of life-forms that are presupposed to have existed in a former geologic age, most often determined by fossil evidence.

Panspermia A phrase coined by Frances Crick representing a belief that aliens planted the seeds of life on earth billions of years ago.

Philosophy The study of principles having to do with being, knowledge or conduct most often in the fields of metaphysics, ethics and logic.

Polymers Large molecular chains composed of repeating units as in proteins and nucleic acids. Synthetic polymers include plastics and silicones.

Progenitor — (Biology) A biologically related predecessor or forefather. A precursor of later development.

Progressivism — Characterized by the principle that doctrines, constitutions and the Bible are not absolute but must be reinterpreted in light of progress in science, culture and technology.

Protein — Organic molecular compounds that are synthesized by plants and animals. Once they are introduced to water from enzymes, proteins yield amino acids, which are required for all organic life processes.

Rock — A mass of consolidated mineral matter assembled by the action of heat, pressure and/or water.

Igneous Rock: Created by the cooling of molten rock (magma) from volcanoes into a solid form; when above ground is called lava rock.

Metamorphic rock: Rock that is composed of various types of rock including igneous, sedimentary and/or other metamorphic rock including limestone and shale.

Sedimentary rock: Little pieces of sandstone, limestone and shale eroded from higher locations to now form the earth's crust.

Science	Systematic knowledge or study of a body of facts or naturally observed truths.
Scientific	Validated knowledge determined by observation, experimentation, measurement and repetition.
Six-Day Creationist	One who believes in the literal six, 24-hour days of creation as taught in the Bible.
Species	(Biology) A major class or kind of genus (plants or animals) bearing some resemblance to one another and able to reproduce or breed with one another.
Spontaneous generation	A-biogenesis, the production of a living organism from a non-living entity.
Strata	The plural form of the Latin word stratum, referring to the observed layering of rocks.
Supernatural	Beyond that which is explained by natural law, as having the characteristics of God or a deity.
Synthesize	To combine different elements into one entity. (Chemistry) to form a more complex substance from simpler elements or compounds.
Tectonics	Pertaining to the formation of the earth's crust and the forces or conditions that cause movement and to designate the results of that movement (earthquakes, etc.) as the present structure of the earth's crust or plate.

Theist	One who believes in one true God of the universe or the existence of a God.
Theory	A coherent group of principles or explanations whose status is not proven and still open to conjecture. Often understood as a hypothesis put forth as a possible explanation.
Transitional	Change or passage from one state of existence to another. Not lasting or permanent.
Uniformitarianism	Derived from a theory suggesting that early geologic processes were the same as those observed today and are without variation throughout past ages.
Unitarianism	Began in 16th-century England as an anti-Trinitarian movement, rejecting the Holy Spirit and Christ as the Son of God. It denies the inspiration of Scripture and claims that church doctrine is the invention of man. Salvation is based on the character of a man and one's understanding of life.
Universe	The universe as such is not referred to in Scripture, though it is understood by the phrase "heavens and earth" as in Genesis 1:1. Most scientists today would include all known as well as supposed objects throughout space—all matter that is measured by time and space.
Zoology	Related to the biological study of animals.

Acknowledgments

One of the greatest joys of our Christian life is the relationships we share with other Christians. During the course of writing this book God blessed me with the aid of several very helpful friends, most of whom were far more science-oriented than I. Their help was invaluable to me.

Jason Thomas, an anesthesiologist with a master's degree in medical science and physical therapy, helped me to understand the molecular complexity of the human body.

Tye Swift, with 50 years of electrical engineering experience in deep-tissue stimulation and electrically sterilized air systems, made me aware of the body's nervous system and its electrical energy circuitry.

Ron Edwards, with a Ph.D. in organic and nuclear chemistry, helped me to refine both literary presentation as well as scientific explanations especially in the field of thermodynamics. Having published several books himself, Ron was able to make some very helpful suggestions to me.

Dr. Henry Blackaby graciously agreed to read my manuscript. I was fortunate enough to share a church pew with Dr. Blackaby for several years at the First Baptist Church in Jonesboro, GA. His humble, Christ-like spirit as well as his spiritually insightful writings have blessed and encouraged me in ways that I could never fully express. I pray that God will allow Henry many more years of faithful service to Him.

Lynn Kaatz, a highly talented professional artist who now lives in St. Augustine, took my scribbled cover suggestion and brought it to life. I hope you enjoy the humor we intended.

Bibliography

Ashton, John F., editor. "In Six Days." Master Books.

Barker, Kenneth L., General Editor. "Zondervan NIV Study Bible New International Version" Published by Zondervan Corporation.

Clark, Ronald W. "The Survival of Charles Darwin." Random House.

Collins, Francis S. "The Language of God." Free Press.

Coulter, Ann. "Godless: The Church of Liberalism." Crown Publishing Group, Random House.

Crick, Francis. "What Mad Pursuit: A Personal View of Scientific Discovery." Basic Books, Inc.

Darwin, Robert Charles. "On the Origin of Species by Means of Natural Selection, Or the Preservation of Favoured Races in the Struggle for Life." London: John Murray, 1859. Republished by Barnes & Noble Books.

DeYoung, Dr. Don. "Thousands . . . Not Billions." Master Books.

Geisler, Norman L. and Turek, Frank. "I Don't Have Enough Faith to Be An Atheist." Crossway Books.

Ham, Ken. "The Lie: Evolution." Master Books.

Ham, Ken. "The New Answers Book 1." Master Books.

Holman. Illustrated Bible Dictionary. Holman Bible Puslishers.

Hovind, Kent Dr. "Creation Seminar Series." CSE Ministry.

Kohlenberger, John R.III. "The Interlinear NIV Hebrew – English Old Testament." Zondervan Publishing House.

Lewis, C. S. "Mere Christianity." Touchstone, Macmillian Publishing.

Lewis, C. S. "The Collected Works of C. S. Lewis." Inspirational Press, BBS Publishing Corp.

MacDonald, William. "Believers Bible Commentary." Thomas Nelson Publishers.

Morris, John, D., President. "Acts and Facts." Institute for Creation Research.

Morris, Henry M. III. "Exploring the Evidence for Creation." Institute for Creation Research.

Morris, Henry M. III & Gary E. Parker. "What is Creation Science?" Master Books.

"Nelson's illustrated Bible Dictionary." Guideposts.

Parker, Steve. "The Human Body Book." DK Publishing.

"Rose Book of Bible Charts, Maps and Time Lines." Rose Publishing, Inc.

The Random House College Dictionary. Random House, Inc.

Wikipedia: The Free Encyclopedia. Wikipedia Web Site: http://www.wikipedia.org.

Breinigsville, PA USA
02 February 2010
231738BV00001B/2/P